Audio Engineering 101

Audio Engineering 101
A Beginner's Guide to Music Production

Timothy A. Dittmar

AMSTERDAM • BOSTON • HEIDELBERG • LONDON
NEW YORK • OXFORD • PARIS • SAN DIEGO
SAN FRANCISCO • SINGAPORE • SYDNEY • TOKYO

Focal Press is an imprint of Elsevier

Focal Press is an imprint of Elsevier
225 Wyman Street, Waltham, MA 02451
The Boulevard, Langford Lane, Kidlington, Oxford, OX5 1GB, UK

Notices
Knowledge and best practice in this field are constantly changing. As new research and experience broaden our understanding, changes in research methods, professional practices, or medical treatment may become necessary.

Practitioners and researchers must always rely on their own experience and knowledge in evaluating and using any information, methods, compounds, or experiments described herein. In using such information or methods they should be mindful of their own safety and the safety of others, including parties for whom they have a professional responsibility.

To the fullest extent of the law, neither the Publisher nor the authors, contributors, or editors, assume any liability for any injury and/or damage to persons or property as a matter of products liability, negligence or otherwise, or from any use or operation of any methods, products, instructions, or ideas contained in the material herein.

Library of Congress Cataloging-in-Publication Data
Application submitted.

British Library Cataloguing-in-Publication Data
A catalogue record for this book is available from the British Library.

ISBN: 978-0-240-81915-0

For information on all Focal Press publications
visit our website at *www.elsevierdirect.com*

11 12 13 14   5 4 3 2 1

Printed in the United States of America

Typeset by: diacriTech, Chennai, India

Working together to grow
libraries in developing countries

www.elsevier.com | www.bookaid.org | www.sabre.org

ELSEVIER   BOOK AID International   Sabre Foundation

Dedication

This book is dedicated to my mom, Jane Dittmar. She encouraged me to be whatever I had passion to be, while inspiring me to create a back-up plan in case my rock drum fantasies never transpired.

Contents

Acknowledgments

Thanks to:

Geoffrey Schulman for getting me into this teaching mess and seeing me as more than a longhaired rock drummer. Without being gently nudged into becoming a professor and getting a real job, this book would probably not exist.

Anderson Bracht for providing inspiring illustrations that kept me smiling.

Catharine Steers, Focal Press, and Elsevier for believing in my book.

Carlin Reagan for making me appear somewhat intelligent and providing mad editing skills.

Terri Dittmar for her feedback, editing, musical contributions, patience, friendship, and years of love and support.

Fred Remmert for teaching me that first lesson and many other audio engineering basics.

All my audio engineering friends who contributed their time and ideas: Meason Wiley, Kurtis Machler, John Harvey, Mary Podio, Andre Moran, and Pete Johnson.

Andrew Miller for producing the *Audio Engineering 101* demonstration videos.

Landry Gideon for being the talent in the *Audio Engineering 101* videos.

My former students who were the guinea pigs for much of the material in the book.

All the audio engineers who contributed to the FAQs section. Much appreciation!

Kathleen Maus and Jack Dittmar for letting me steal your 45s to broadcast K-TIM.

Dave Dittmar for letting me play in his band when I was 10 years old, teaching me the basics of electronics, and how to work a PA.

Susan Dittmar for helping kick-start my audio engineering career. Without her, I wouldn't be writing this book.

When I was in high school, my mom asked me, "Tim, what do you want to do with your life?" I believe I said something like, "play drums and surf." My mom was pretty cool about the whole thing, but I knew my ambitions did little to impress her. Soon after graduating high school, I enrolled at a local college. There I discovered I could obtain a degree in a field related to my interests, Radio TV Production. I loved music, but the idea of recording and working with musicians for a living seemed to be an unattainable dream! I moved to Austin, TX, in the late 80s to play drums and complete my Radio-TV-Film (RTF) degree at The University of Texas. I soon found myself at Cedar Creek Recording studio playing drums on a new wave demo. It was my first experience in a commercial recording studio and I was blown away. The 2" tape machine, the array of mics, the recording console, and the reverb! Oh, the Lexicon 224 reverb on my snare. I was hooked. That first night in the studio I was so wound up and boisterous that the engineer actually told me to shut up. The engineer's comments taught me the first lesson in studio etiquette. Don't talk, listen! Even with this reprimand, I was given the opportunity to become an intern at CCR and spent the next ten years engineering records.

Looking back, while I had only a basic foundation in recording, I remember feeling technically inept. I had to rely on a variety of strengths to compensate for my lack of technical expertise. Some of these strengths included showing up on time and having a good attitude and a willingness to listen. I was also very self-motivated and driven, had decent people skills, and was willing to struggle and work hard for little money while learning the craft. I also knew a few bands and musicians, and could bring in potential business. It took years before my technical skill level matched these intangible strengths. My point is, if you are a beginner, do not be discouraged. We all have to start somewhere.

Preface

Audio engineering is layered with many disciplines such as acoustics, physics, technology, art, psychology, electronics, and music. Each layer provides a new set of questions, answers, theories, and concepts. You soon realize that you will never know it all. This guide encompasses many aspects of audio engineering and includes a dose of reality. This is a hard business, but if you are willing to put forth the time and effort, you may end up with a job you love.

The goal of this book is to explain audio engineering and music production in an easy-to-understand guide. After ten years of teaching Audio Engineering courses at Austin Community College, two years of lecturing at The University of Texas, and twenty-five years of engineering and producing records, I decided to create a guide that draws on the lessons and experiences that have proved to be successful with both students and clients. As a Professor, I have unique insight into a beginner's ability to retain technical information. Many audio engineering books are simply too overwhelming for those being introduced to recording and music production. This is a recurring point made by many students and is one of the inspirations for this book. *Audio Engineering 101* explains intangible concepts that can make a recording better, such as understanding body language, creating a good vibe, and people skills. Much of your business will be generated by word-of-mouth, so these are important skills. In addition, the book highlights what to expect with internships, how to create a productive recording space, and an overview of what jobs are available to audio engineers. You will also find a handy guide dedicated to microphones and their uses. This is a great section for a true beginner or for a hobbyist wanting to learn microphone basics. *Audio Engineering 101* includes FAQs (frequently asked questions) answered by a diverse group of professional recording engineers from around the country. Questions are answered by experienced Pros: what is the first mic you should buy or how you can get your foot in the door.

You can't learn everything about music production in a semester of school or even by getting a degree in the subject. Becoming proficient in music production may take many years. Experience is one of the most valued aspects of the profession and is gained by creating thousands of mixes, both good and bad, learning from your mistakes, and continually honing the craft. This is one of the coolest jobs you could ever have, but it won't be easy becoming a true professional. Even if you decide not to become an audio engineer, this book will take some of the mystery and intimidation out of the studio and the recording process.

CHAPTER 1

What Is Sound? Seven Important Characteristics

Learning the craft of audio engineering is like learning a foreign language. A new language may be intimidating and difficult at first, but with time and dedication, a vocabulary is built. Soon words turn into phrases and phrases turn into full sentences. This chapter will cover details of a sound wave and explore some of the language of audio. You will be fluent in no time!

WHAT IS SOUND?

Sound is a vibration or a series of vibrations that move through the air. Anything that creates the vibrations, or waves, is referred to as the source. The source can be a string, a bell, a voice, or anything that generates a vibration within our hearing range.

Imagine dropping a stone in water. The stone (source) will create a series of ripples in the water. The ripples (waves) are created by areas of dense molecules that are being pushed together as sparse molecules expand, thus creating flatter areas. Sound travels just like this, by compression and rarefaction. Compression is the area where dense molecules are pushed together and rarefaction is the area where fewer molecules are pulled apart, or expanded, in the wave. The compression area is higher in pressure and the rarefaction area is lower in pressure.

This chapter deals with the seven characteristics of a sound wave, such as amplitude, frequency, phase, velocity, wavelength, harmonics, and envelope. Understanding these characteristics is essential to make a decent recording, become a competent mix engineer, and generally increase your knowledge about audio. Although a typical sound is more complex than a simple sine wave, the sine wave is often used to illustrate a sound wave and its seven characteristics.

FIGURE 1.1

SEVEN CHARACTERISTICS OF SOUND

You may already know about amplitude and frequency. If you have ever adjusted the tone on your amp or stereo, then you have turned up or down the "amplitude" or a "frequency" or range of frequencies. It is necessary to understand these two important sound wave characteristics, as they are important building blocks in audio engineering. Two other characteristics of sound help humans identify one sound from another: harmonics and envelope. The remaining three

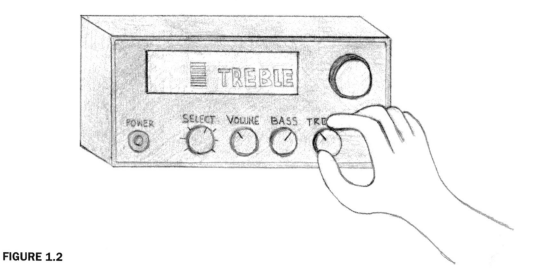

FIGURE 1.2

characteristics of sound are velocity, wavelength, and phase. These characteristics identify how fast a sound wave travels, the physical length of a completed cycle, and the phase of the sound wave.

Amplitude

Amplitude is associated with the height of a sound wave and is related to volume.

When a stereo, amp, or television's volume is turned up or down, the amplitude of the sound being projected is increased or decreased. Loud sounds have higher amplitudes while quiet sounds have lower amplitudes. The greater the amplitude of a sound the greater the sound pressure level.

Amplitude is measured in decibels (dB). Most people can recognize about a 3 dB change in amplitude. A trained ear can recognize even smaller amplitude changes. An increase in amplitude is usually expressed as a "boost" and a decrease in amplitude is often expressed as a "cut." The word volume is often substituted for amplitude. An audio engineer may say, "boost that 3 dB" or "cut that 3 dB." When amplitude is written out, it is expressed with a positive sign such as +3 dB or a negative sign such as −3 dB.

FIGURE 1.3

Here are some common activities and their corresponding decibel levels:

    0 dB – near silence
    40–50 dB – room ambience
    50–60 dB – whisper

60–75 dB – typical conversation

80–85 dB – a blender, optimum level to monitor sound according to the *Fletcher–Munson curve*

90 dB – factory noise, regular exposure can cause hearing damage

100 dB – baby crying

110 dB – leaf blower, car horn

120 dB – threshold of pain, can cause hearing damage

140 dB – snare drum played hard from about 1'

150–160 dB – jet engine

As you can see, in our daily lives, we are constantly confronted with amplitude levels between 0 dB and about 160 dB. Most people listen to music between 70 dB (on the quiet side) and 100 dB (on the loud side). Appendix A covers dBs in more detail.

Frequency

The amount of cycles per second (cps) created by a sound wave is commonly referred to as the frequency. If you are a musician, you may have tuned your instrument to A/440. Here, "440" is the frequency of a sound wave. Unlike amplitude, which is measured in decibels, frequency is measured in hertz (Hz), named after the German physicist, Heinrich Hertz. The average human hearing range is from 20 to 20,000 Hz. Typically, once 1000 cycles per second

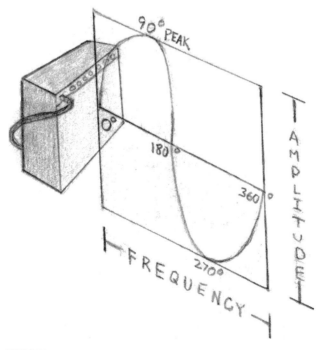

FIGURE 1.4

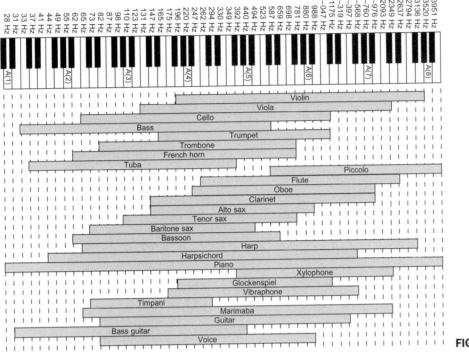

FIGURE 1.5

is reached, the frequency is referred in kilohertz (kHz), i.e., 1000 Hz = 1 kHz, 2000 Hz = 2 kHz, and 3000 Hz = 3 kHz. Frequency is related to the pitch of a sound. Here is a handy chart to help identify the frequency ranges of various instruments and how the keys of a piano relate to frequency. The first note on a piano is A, which is 27.5 Hz. Have you ever turned up the bass or treble on your car stereo? If so, you are boosting or cutting the amplitude of a frequency or range of frequencies. This is known as equalization (EQ), a vital aspect of audio production.

Each frequency range has distinct characteristics, and some common terms can help you to identify them. I will go into further detail throughout the book, but let's start here:

Frequency is often divided into three ranges:

AUDIO CLIP 1.0

Low or bass frequencies are generally between *20 and 200 Hz*. These frequencies are omnidirectional, provide power, make things sound bigger, and can be destructive if too much is present in a *mix*.

Mid, or midrange, frequencies are generally between *200 Hz and 5 kHz*. This is the range within which we hear the best. These frequencies are more directional than bass frequencies and can make a sound appear "in your face," or add attack and edge. Less midrange can sound mellow, dark, or distant. Too much exposure can cause ear fatigue.

High or treble frequencies are generally between *5 and 20 kHz* and are extremely directional. Boosting in this range makes sounds airy, bright, shiny, or thinner. This range contains the weakest energy of all the frequency ranges. High frequencies can add presence to a sound without the added ear fatigue. A lack of high frequencies will result in a darker, more distant, and possibly muddy mix or sound.

Midrange is the most heavily represented frequency range in music. It is often broken down into three additional areas:

Low-mids, from around 200 to 700 Hz darker, hollow tones
Mid-mids, from 700 to 2 kHz more aggressive "live" tones
High-mids or upper-mids, from 2 to 5 kHz brighter, present tones

This chart may come in handy when you are learning how to describe a particular sound or when you are mixing. These are general areas and are covered in detail in Chapter 3.

FIGURE 1.6

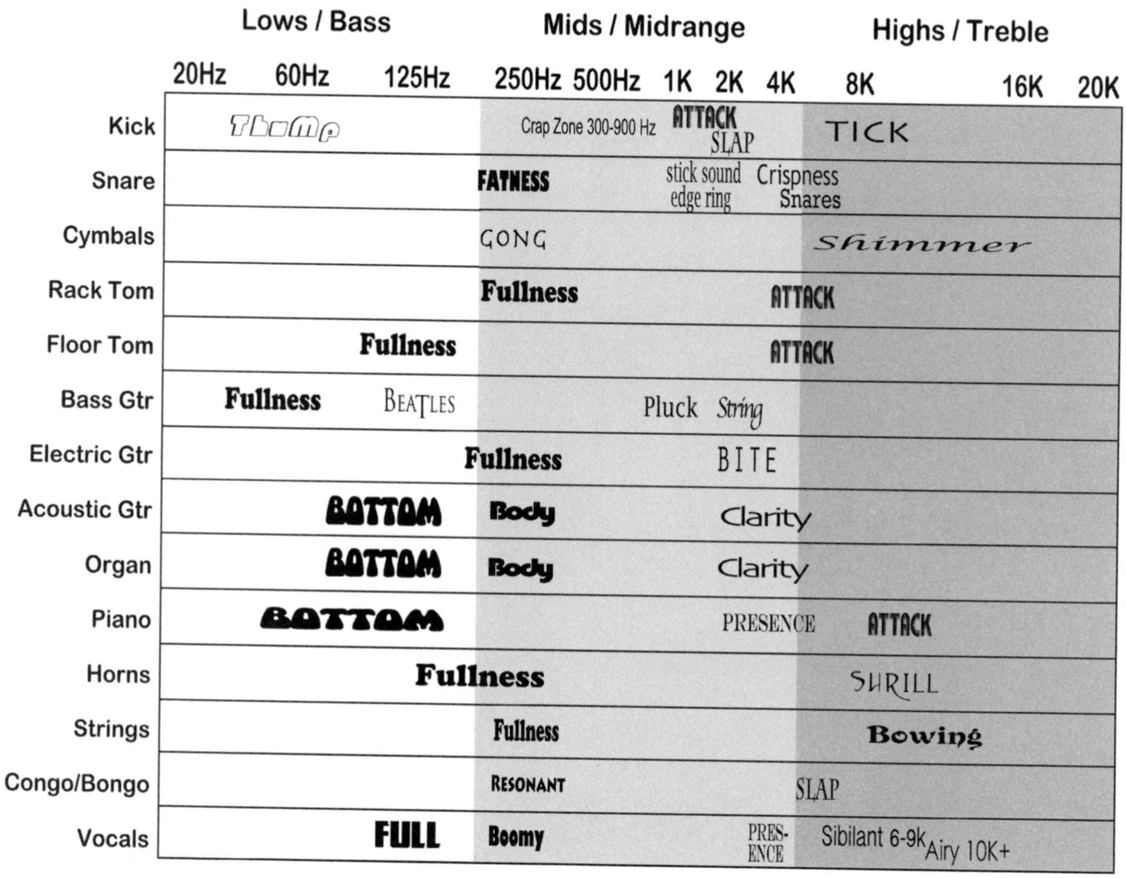

	Lows / Bass			Mids / Midrange					Highs / Treble		
	20Hz	60Hz	125Hz	250Hz	500Hz	1K	2K	4K	8K	16K	20K
Kick	Thump			Crap Zone 300-900 Hz			ATTACK SLAP		TICK		
Snare				FATNESS			stick sound edge ring	Crispness Snares			
Cymbals				GONG					Shimmer		
Rack Tom				Fullness				ATTACK			
Floor Tom			Fullness					ATTACK			
Bass Gtr	Fullness		BEATLES			Pluck	String				
Electric Gtr				Fullness			BITE				
Acoustic Gtr		BOTTOM		Body			Clarity				
Organ		BOTTOM		Body			Clarity				
Piano		BOTTOM						PRESENCE	ATTACK		
Horns			Fullness						SHRILL		
Strings				Fullness					Bowing		
Congo/Bongo				RESONANT				SLAP			
Vocals			FULL	Boomy				PRES-ENCE	Sibilant 6-9k Airy 10K+		

Phase

Phase designates a point in a sound wave's cycle and is also related to frequency, see Fig. 1.3. It is measured in degrees and is used to measure the time relationship between two or more sine waves.

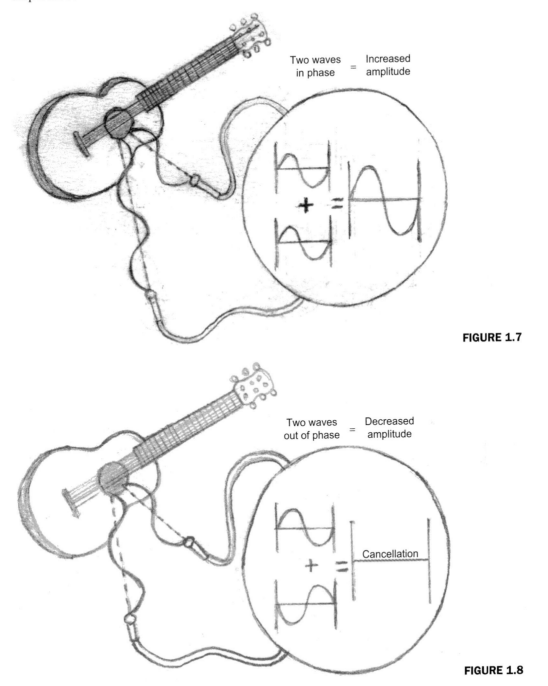

Two waves in phase = Increased amplitude

Two waves out of phase = Decreased amplitude

Cancellation

FIGURE 1.7

FIGURE 1.8

As you can see from Figs 1.6 and 1.7, when two sound waves are in phase, the result is increased amplitude. When they are 180 degrees out of phase, they can completely cancel each other resulting in little or no sound. This concept is used in many modern devices, such as noise-cancelling headphones or expensive car mufflers, to eliminate the outside sound or engine noise. However, sound is not always completely in or out of phase. Sounds can be out of phase by any number of degrees, ranging from 1 to 359. Phase issues can make some frequencies louder and others quieter. Often a room's acoustics create these areas of cuts and boosts in the frequency spectrum. These cancellations and amplitude increases influence the way a room is going to sound. Standing waves and comb filtering are often the result of these phase interferences. Phase is also very important to keep in mind when stereo miking and when using multiple mics on an intended source. When listening in a typical stereo environment, a sound may be completely out of phase and go unnoticed unless the phase is checked.

▲◉ TIP

Some tips to check phase:

> Mono button
> Phase flip (polarity)
> Phase meter

Phase issues can be exposed when a mix or a sound is checked in mono. One of the first records I mixed was a new wave record with thick delays and effects. I was mixing a particular track for a music video. The studio where I was working at that time had a small TV with a mono speaker. I would patch mixes into the TV in order to hear the mixes in mono. This would expose any existing phase issues and instrument imbalances. I patched into that TV after completing what the band and I thought was a pretty good mix, and the vocals and keyboards almost completely disappeared! Imagine if I hadn't checked the phase in mono. The video would have been aired and there would have been no vocals. I can honestly say after that experience that the mono button became one of my go to buttons on the recording console. Many live music venues and dance clubs' PAs and speaker systems are set-up in a mono configuration to get more power. What would happen if one of your out-of-phase mixes were played in a club? It would be less than impressive. Always check your mixes in mono!

Velocity

Velocity is the speed at which sound travels. Sound travels about 1130 ft per second at 68 degrees Fahrenheit (344 m/s at 20°C). The speed at which sound travels is dependent on temperature. For example, sound will travel faster at higher temperatures and slower at lower temperatures, knowing that the velocity of sound can come in handy when calculating a *standing wave* or working with live sound.

Wavelength

Wavelength is the length of the sound wave from one peak to the next. Consider the wavelength to be one compression and rarefaction of a sound wave. In determining the wavelength, the speed of sound and divide it by the frequency. This will identify the length between these two peaks.

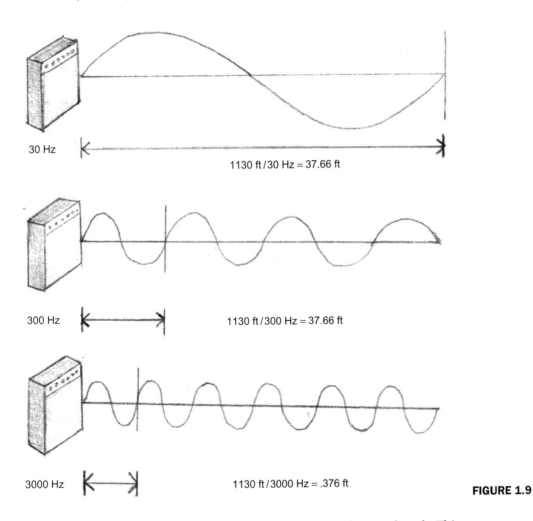

30 Hz

1130 ft / 30 Hz = 37.66 ft

300 Hz

1130 ft / 300 Hz = 37.66 ft

3000 Hz

1130 ft / 3000 Hz = .376 ft

FIGURE 1.9

As seen in the figures, the lower the frequency the longer the wavelength. This demonstrates the power and energy that low end creates as a result of a longer wavelength. High frequencies are much smaller in length resulting in a weaker form of energy that is highly directional.

⚠ TIP

The Ruben's Tube is a great visual example of compression, rarefaction, frequency, and wavelength. Look up the Ruben's tube built by The Naked Scientists on their Garage Science blog: http://www.thenakedscientists.com/HTML/content/kitchenscience/garage-science/exp/rubens-tube/.[1]

Unlike other sound wave characteristics previously discussed, harmonics and envelope help humans differentiate between one instrument or sound from the other.

Harmonics

The richness and character of a musical note is often found within the harmonics. Harmonics are commonly referred to as "timbre." Every instrument has a fundamental frequency, referred to as the fundamental, and harmonics associated with it. On an oscilloscope, the fundamental shows up as a pure sine wave, as seen in the Ruben's Tube video; however, sound is much more complex. Most sounds contain more information in addition to the fundamental. In music, instruments have their own musical makeup of a fundamental plus additional harmonics unique to that instrument. This is how we can distinguish a bass guitar from a tuba, a French horn from a violin, or any two sounds when the same note at the same volume is played. Instruments that sound smoother, like a flute, have less-harmonic information and the fundamental note is more apparent in the sound. Instruments that sound edgier, like a trumpet, tend to have more harmonics in the sound with decreased emphasis on the fundamental.

If you were to play a low E on the bass guitar, known as E1, the fundamental note would be about 41 Hz. You can figure out the harmonics by simply multiplying the fundamental times 2, 3, 4, etc.

The fundamental note E1 = 41 Hz.
The second harmonic would be 82 Hz (41 × 2).
The third harmonic would be 123 Hz (41 × 3).
The fourth harmonic would be 164 Hz (41 × 4).

It is a common practice among engineers to bring out a sound by boosting the harmonics instead of boosting the fundamental. For instance, if the goal is to hear more bass, boosting 900 Hz may

FIGURE 1.10

[1]"Rubens' Tube-waves of fire." www.thenakedscientists.com. The Naked Scientists, n.d. Retrieved from The Naked Scientists, http://www.thenakedscientists.com/HTML/content/kitchenscience/garage-science/exp/rubens-tube/ (June 2011).

bring out the neck, or fret board, of the instrument and make the note pop out of the mix. The result is more apparent in bass, without the addition of destructive low end to the instrument.

Additionally, harmonics are divided into evens and odds. Even harmonics are smoother and can make the listener feel comfortable, whereas odd harmonics often make the listener feel edgy. Many engineers and musicians use this knowledge when seeking out microphone preamps, amplifiers, and other musical equipment containing vacuum tubes. These tubes create even distortion harmonics that are pleasing to the ear and odd distortion harmonics that generate more edge and grit.

FIGURE 1.11

🔺 TIP

Taking a music fundamentals class or studying music theory can definitely benefit you as an audio engineer. These classes and concepts can help you develop a well-rounded background and better understanding of music. You can never know too much in this field!

The more you know, the easier time you will have communicating effectively with skilled musicians. If you are able to speak intelligently, they are more likely to be comfortable working with you and putting their trust in you. The more skills you possess the better your chance for success.

Envelope

Like harmonic content, the envelope helps the listener distinguish one instrument or voice from the other. The envelope contains four distinct characteristics: attack, decay, sustain, and release.

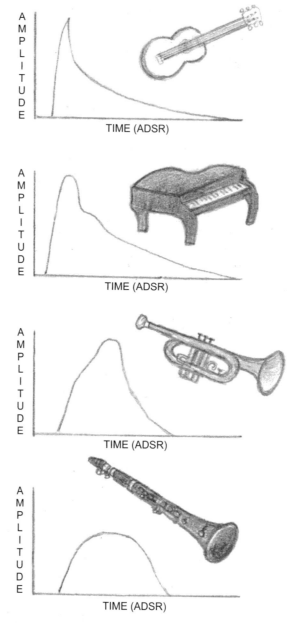

FIGURE 1.12

Attack is the first point of a note or sounds envelope. It is identified as the area that rises from silence to its peak volume.

Decay is the next area of the envelope that goes from the peak to a medium level of decline.

Sustain identifies the portion of the envelope that is constant in the declining stage.

Release identifies the last point in the envelope where the sound returns back to silence.

A percussive instrument has a very quick attack, reaching the note instantly upon striking. With woodwinds, brass, and reed instruments, no matter how quickly the note is played, it will never reach the note as fast as striking a drum.

OTHER PERIODIC WAVEFORM TYPES

Waveform defines the size and shape of a sound wave. Up to this point, a simple sine wave has been used to illustrate sound. Sound can come in different waveforms, other than a sine wave. Other common waveforms include triangle, square, and sawtooth waves. Each waveform has its own sound and characteristics and each may be used for different applications.

A triangle wave looks like a triangle when viewed on an oscilloscope, a square wave appears as a square, and a sawtooth wave appears as a sawtooth.

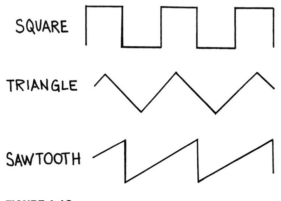

SQUARE

TRIANGLE

SAWTOOTH

FIGURE 1.13

A square wave is typically associated with digital audio. A square wave's sound is often described as hollow and contains the fundamental note plus the odd harmonics. These harmonics gradually decrease in amplitude as we go higher in the frequency range.

A triangle wave is similar to a square wave in that it also contains only the fundamental plus the odd harmonics. It is a kind of a cross between a sine wave and a square wave. One main difference is that the higher frequencies harmonics are even lower in amplitude than those of square waves. This results in a less harsh sound and is often used in synthesis.

A sawtooth wave contains both the even and the odd harmonics of the fundamental. Its sound is harsh and clear. Sawtooth waveforms are best known for their use in synthesizers and are often used for bowed string sounds.

AUDIO CLIP 1.1

NOISE

Noise is any unwanted sound that is usually non-repeating. Noise is a hum, a hiss, or a variety of extraneous sounds that accompany a sound wave when it is mixed or recorded. Noise comes from a variety of sources besides the instrument, such as an air conditioner, fluorescent lights, or outside traffic.

One way to express quality of sound is by using the Signal-to-Noise Ratio, shortened S/N. This ratio compares the amount of the desired signal with the amount of unwanted signal that accompanies it. A high-quality sound will have significantly more signal (desired sound) than noise (undesired sound).

Distortion, unlike noise, is caused by setting or recording levels too hot, pushing vacuum tubes, or by bad electronics. When needed, adding it can be an effective way to make a sound dirty, more aggressive, or in your face.

Headroom is the maximum amount a signal can be turned up or amplified without distortion. As an audio engineer you should be aware that audio devices have different amounts of headroom. Make sure you allow for plenty of headroom when setting audio signal levels. If you don't, a loud spike of sound may ruin a take. Analog level settings can exceed zero, while digital cannot. Level settings will be discussed in Chapters 8 and 11.

In this chapter, we learned about seven key sound wave characteristics: amplitude, frequency, velocity, wavelength, phase, harmonics, and envelope. Distinguishing between a decibel and a hertz, or a low-frequency and a high-frequency sound will be very important with music production. Having a basic grasp of these terms will help create a solid foundation in audio engineering.

CHAPTER 2

How to Listen. Remember When Your Parents Told You to Listen? Well, You Should Have Listened!

This chapter examines the ear and how sound is interpreted. We will also discuss other important skills and ideas like analyzing music recordings, the frequency pyramid, and the 3D reference ball. Understanding human hearing and learning how to better interpret frequencies will result in an increased ability to manipulate sound.

HOW THE EAR WORKS

The Ear

Like a microphone, the ear is a transducer. Our ears convert acoustic sound vibrations into mechanical energy which is then sent to our brains as electrical impulses. The human ear is made up of the outer, middle, and inner ear. The bones in our middle ear help amplify sound, while the inner ear's muscles help protect it from loud or sudden volume changes.

In the previous chapter, frequency was divided into three areas: the bass, or lows, from 20 to 200 Hz; midrange frequencies from about 200 Hz to 5 kHz; and treble, or highs, from 5 to 20 kHz. Most humans hear best in the following order: midrange, highs, and then lows. However, as volume increases, the different frequency ranges are heard more equally. According to the Fletcher–Munson curve, most humans will hear all frequency ranges equally between 80 and 85 dB. It is important to understand what frequencies will be heard at any given volume. When you are in the studio monitor mixes at both lower and higher volumes to ensure that the recording sounds good when the music is played quietly or at full volume.

TIP

Turn on your stereo, iPod, or personal listening device. Listen at a low volume and note which instruments are heard well. You are likely to hear the vocals, snare drum, and other midrange tones. Next, increase the volume and notice how the bass range fills up the overall sound.

Around 95 dB, the ear's natural defense is to limit the volume of the sound. At volumes more than 95 dB, the brain interprets bass sounds as flatter and treble sounds as sharper in pitch. This is one of the many reasons not to blast headphones, especially for singers! If you have ever been to a loud concert, you probably recognize that your ears adjust to the louder volume. Our ears actually limit these louder sounds in an attempt to prevent damage. As previously mentioned, humans are unable to hear sound accurately at louder volumes. Limit the time you spend listening to loud volumes. Consider taking short breaks throughout a session to allow the ears time to recover. How can you expect a great result if you cannot hear accurately due to ear fatigue?

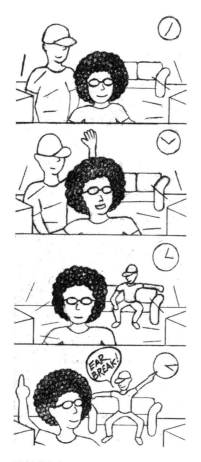

FIGURE 2.1

Extended exposure to loud sound can lead to permanent hearing loss. Keep earplugs handy for loud concerts and other events with loud volumes. If you are practicing with your rock band in the garage, wear earplugs! They may be annoying at first, but you will get used to them. There are a variety of different types of earplugs, at different price ranges. The typical foam earplugs reduce volume and more of the mid and higher frequencies. These are intended more for industrial use than for music application. These are effective in reducing extreme volumes, but can severely compromise the tone of a sound (Fig. 2.2).

Some companies make earplugs specifically suited for use when listening to and working with music and are available at local music stores or online. These earplugs decrease the amplitude, but retain the frequency range (Fig. 2.3).

Also, an Audiologist can provide custom earplugs that are suited for more critical listening and are molded to fit your ear. The professional molded earplugs can include inserts for −10, −15, and −20 dB attenuators along with other filters (Fig. 2.4).

Links for audio earplugs:

www.hearos.com
www.earplugstore.stores.yahoo.net/
http://www.hear-more.com/musician.htm

Age and gender contribute to a person's ability to hear sound. As we get older, many of us will not be able to hear frequencies much above 16 kHz, while some of us may not be able to hear much above 13 kHz. The legendary George Martin can only hear up to about 12 kHz. Most of our "perfect" hearing years occur before our 25th birthday. In terms of gender and hearing, women tend to hear higher frequencies better than men. The inner ear hairs (organ of corti) and cochlea are stiffer in women, resulting in a more sensitive frequency response. Statistics show that men have more hearing deficiencies, and women's hearing deteriorates at a slower pace in comparison with men.

FIGURE 2.2   **FIGURE 2.3**   **FIGURE 2.4**

To learn more about your own ears, you can get your hearing checked by an Audiologist. To check your hearing for free, go to www.digital-recordings.com/, where there are several great online tests.

Ears are sensitive transducers with a non-linear frequency response. Frequencies that humans hear best are between 1 and 4 kHz, and at lower volumes, the ear doesn't respond well to bass frequencies. At lower volumes, the three small bones in the middle ear help amplify quieter sounds so that we can hear them better. To protect your ears at louder volumes, insert earplugs and enjoy.

Also worth considering is the resonant frequency of the ear canal, which is roughly 3 kHz in an average adult ear. This resonant frequency means we humans are more sensitive to that particular frequency. One interesting point is that a newborn baby's cry is also roughly 3 kHz.

⚠ TIP

To ensure the mix is not bass heavy or bass light, monitor sound between 80 and 85 dB. Remember, this is the volume where frequency ranges are heard more equally. To get to know what 80–85 dB sounds like, you can purchase a dB or a sound pressure meter at your local electronics store, music store, or find one online. There are also multiple phone apps that offer different types of sound pressure meters.

Direct, Early Reflections, Reverberation

Sound can be divided into three successively occurring categories that arrive at the ears in the following order: direct path, early reflections, and reverberation.

The direct path is the quickest path to the listener. It helps identify the location where a sound is coming. As binaural creatures, we use two ears with our head acting as an object blocking the sound between each ear, to determine the direction of a sound. If a sound reaches our left ear first, we recognize the sound as originating from the left side. The same applies to the right side. If a sound reaches both ears at the same time, the sound originates either directly in front and center or directly behind us.

Early reflections occur immediately after the direct sound. These reflections clue us in to the surface type (wood, tile, carpet, etc.) around a particular sound. We hear early reflections from all directions fairly equally. Early reflections appearing from the sides make a sound appear wider and more spacious. Our brains have a difficult time interpreting early reflections under about 20 ms. This psychoacoustic phenomenon is known as the Haas effect. We can actually use this to fatten up a sound. For example, try putting a 6 ms delay on the bass, voice, or guitar. Pan the dry, or unprocessed, signal left and the delayed, or wet, signal right. This will stretch the signal between the two speakers providing a single thick sound. The signal will not be heard as two individual sounds.

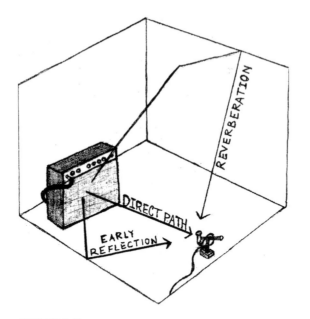

FIGURE 2.5

FIGURE 2.6

Reverberation is the last component of sound that is processed by the brain. This is how the size of the space where the sound originated is determined. Reverberation is the sound decay or "wash" left after the initial sound and early reflections. The longer it takes for the wash or reverb to go away, the larger is the space. The words "environment" and "space" may also be used to represent the word reverberation.

Direct path, early reflections, and reverberation are not necessarily sound waves you will hear independently of one another. The time differences that can occur between each one are so minute that they are typically interpreted as a single sound. Our brain processes a sound and provides us clues to its direction, the surface material around it, and the size of the space or environment, even though we think we are hearing a single sound. We can use this information to recreate spaces, to better evaluate sound, and to add depth and dimension to our recordings.

EAR TRAINING TOOLS AND TECHNIQUES
3D Reference Ball

1 Explaining the 3D Ball

Audio engineers work with intangible tools: sounds, pitches, and instruments. It can be difficult to manipulate these invisible elements into a well-rounded mix. To help create a full mix, when monitoring or recording, imagine a 3D ball suspended between the two speakers. This gives you a physical reference to work with. In this space, you will control where the sounds are situated – whether be it the front or back, left or right, and top or bottom. The specific controls used to arrange sounds within this 3D reference ball will be discussed in detail in Chapter 6, "Mixing Consoles." Many new engineers do not take advantage of the entire 3D space, the reference ball, thus creating a mix lacking dimension, texture, and clarity.

Different tools are used to manipulate sound in different ways. In order to move sound around in the 3D ball, volume/amplitude, reverberation, also known as environment, panning, and EQ are used. One way to move a sound to the forefront of the 3D space, or toward the listener, is to turn up the volume or amplitude. To move a sound to the background, turn down the amplitude. Likewise, boost or cut midrange to influence where a sound is placed in the ball. Use less midrange to make the sound appear further back and more midrange to make the sound appear upfront and in your face. Another way to move a sound to the front or back of the 3D ball is to apply reverberation (environment). Consider reverb, or environment, as the ambient sound created by a hall, garage, living room, or other acoustic space. Apply less environment or reverb to bring a sound or image toward the listener, or the front of the ball. Use more environment or reverb to push a sound away from the listener, or to the back of the 3D space.

Panning is another tool used to control sound placement within the 3D ball. Panning is applied to move a sound from left to right and is often times described as time on a clock face. An image panned to 3 o'clock will appear to mainly come from the right speaker. An image panned to 9 o'clock will appear mainly from the left speaker. A hard-panned image appears strictly from that speaker side. Panning is one of the most effective ways to separate two like instruments within the 3D space, such as two guitars or backing vocals. Often secondary instruments, such as tambourine, shaker, bells, hi-hats, and cymbals, are panned hard. Bass guitar and kick drum are rarely panned hard left and right because the powerful low-end associated with these instruments is efficiently divided between the two speakers and bass is omnidirectional. Consider them as your anchors. The main vocals are also usually centrally panned to draw the listener's attention (Fig. 2.7).

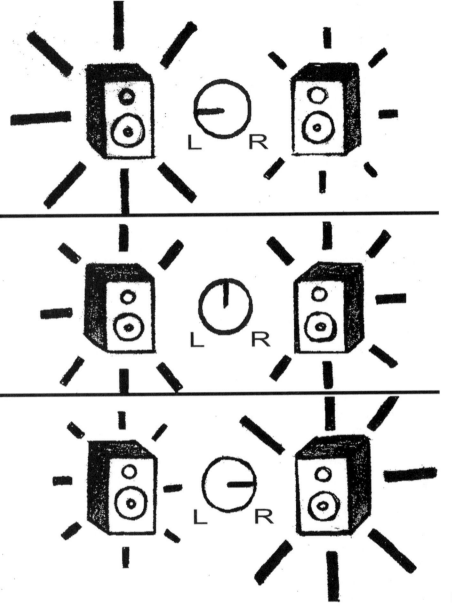

FIGURE 2.7

Finally, adjusting the EQ will move sound up or down in the 3D reference ball. When listening to typical monitors, the high frequencies that emanate from the tweeter will be highly directional and will appear around ear level. The mid and low frequencies will be directed to the woofer at the bottom of the speaker and will sit at the lower portion of the 3D ball. The midrange will hit you in the chest and the lows will surround you and fill up the floor.

Don't be afraid to utilize the entire 3D space. Visualize where each sound will sit in the final mix. One thing you want to avoid is positioning everything in the middle of the 3D space, the equivalent to up front in the mix. This results in a very 1D mix and makes it more difficult for the listener to focus on any one sound. Try to envisage yourself watching the song live and imagine where the instruments/musicians would be on the stage. This image in your mind should help when it comes to positioning within a mix.

▲ TIP

When I am mixing I try to imagine an old school balancing scale. If I pan one guitar right, I want to put something of equal weight on the other side, say another guitar, or trumpet.

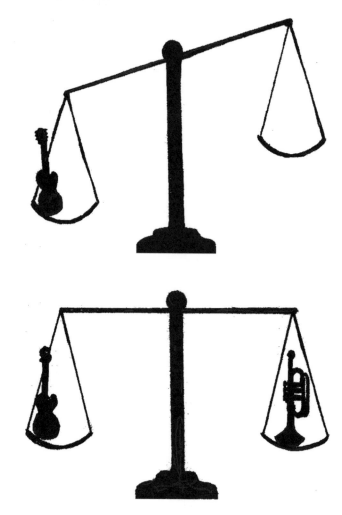

FIGURE 2.8

Frequency Pyramid

Another visualization tool is to consider building a pyramid with the different frequency ranges. The pyramid would be built with low frequencies on the bottom, mid frequencies in the middle, and high frequencies on the top. While recording and mixing, keep this pyramid in mind and make sure you have enough of each frequency range. If the pyramid is lacking a particular frequency range, the mix may sound unbalanced or incomplete.

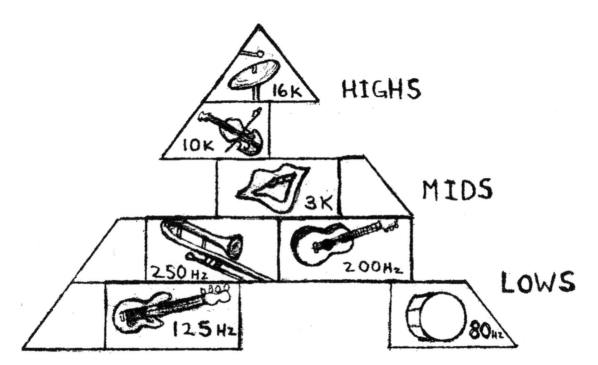

FIGURE 2.9

Selective Hearing

Selective hearing is essential if you want to succeed as an audio engineer or musician. Selective hearing involves focusing on individual components of a sound, the different frequency ranges, and the overall sound. As an audio engineer, you should be able to pick out any instrument in a mix, focus on it, and evaluate the dynamic and tonal relationship it has with the other instruments or sounds.

 TIP

Tips for selective hearing:

- Practice focusing on an individual instrument for the entire song.
- Try to follow every note or beat.

- Repeat this for each instrument or sound in the mix.
- Listen to like instruments, for example, bass and kick drum or two electric guitars. Examine their relationship in the mix.
- Focus on groups of instruments, such as strings, vocals, and rhythm tracks. What are the overall relationships in the mix?

Analyzing Music Recordings

Selective hearing involves being able to pick out an instrument in a recording or mix. Related to selective hearing is the ability to analyze and recreate a sound. Just like a filmmaker watches a film and notices the different camera angles, lighting, and movement of the actors, it will be your job as an audio engineer to identify certain details in a mix or recording. These details may include noticing tones, reverbs and effects, the balance of instruments, and other specific production details. Analyzing music recordings to help mimic a particular sound for a client or to be able to communicate with other audio professionals is essential in music production.

🔺 TIP

Try the following exercise:

1. Listen to a song and identify all the instruments or sounds in the mix.
2. Where are all the sounds placed in the 3D reference ball? Which sounds appear in the center? Which sounds appear on the sides? Do any instruments have a stereo spread?
3. How do the vocals sit in the mix? Are they in the front of the ball or do they blend more with the music?
4. What are the tones of the individual sounds? Are some sounds bright while others are dark or dull? What is the overall tone of the recording?
5. Can you hear the environment around each sound? Do all the instruments appear to be in the same space? What does the overall environment sound like?
6. Is it a "wet" mix or a "dry" mix? A wet mix will have a lot of effects, such as reverb, delay, or echo, and is often used in pop, psychedelic, surf, and reggae music (see Chapter 7). A dry mix will be more organic with little or no effects apparent in the mix. Dry mixes are more common to folk, blues, jazz, classical, and bluegrass music.
7. Finally, and most importantly, how does the recording make you feel? (The recording makes me feel like driving faster, sitting on the couch, smiling, etc.)

You may want to compare two separate recordings, especially when determining the overall tone. You might think that the recording you are analyzing is fairly dark, but compared to what? It is always good to establish some type of reference point between two different instruments or mixes. Comparing one sound against another is often referred to as "A/B."

This chapter discussed how the ear works and interprets sound. If you understand those things, you should be able to evaluate and adjust sound. In music production, it is important to analyze recordings. Selective hearing along with imagining the 3D reference ball and the frequency pyramid will make this task easier. Listen to music and practice these exercises. Auricula has a half-dozen ear training applications that can be found at www.auriculaonline.com. Auricula's Helices is a plug-in that can be purchased for Garage Band and they also offer a free phone app. This is a great tool for learning how to recognize different frequencies and for general ear training. The next chapter will introduce ways to communicate with musicians and provide some additional vocabulary to express sound quality.

CHAPTER 3

EQ Points of Interest. Frequencies Made Easy

EQUALIZATION (EQ) AND FREQUENCY

Equalization, or EQ, can be used to describe the action of equalizing a sound, a control to change the tone, or a reference to the tone of a sound. More than likely you have already equalized something in your life. If you have ever changed the bass or treble settings on your car or home stereo, then you have performed this basic engineering function. In audio production, there are a variety of equalizer controls at your disposal, to change the tone of a recording. Equalizers, also called EQs, are available as standalone rack units, as part of a channel strip, and as software plug-ins.

What actually happens when a sound is equalized? The tone of an overall sound is altered by increasing or decreasing the amplitude of a particular frequency or a range of frequencies, such as bass. Remember the terms frequency and amplitude, found in Chapters 1 and 2? They are two essential elements in understanding audio, especially when we are discussing equalization.

Understanding the different frequency ranges and how to describe them is a necessary skill before you can begin to equalize. It is important to be familiar with specific frequencies and how they are often described and reproduced. This will make it much easier for you, as an engineer, to create or re-create a sound the client may be describing.

Although there are exceptions, most musicians do not communicate using technical terms like "boost 100 Hz 3 dB on my bass." They are more likely to describe something in layman's terms. "I wish my bass sounded 'fatter'," or "My bass sounds too 'thin'." While there is no universal language to describe sound, there are many helpful ways to communicate with musicians who may describe sound quality in their own ways.

In this chapter we will discuss common EQ properties to help you identify frequencies quickly and communication tips for talking with musicians who often speak in layman's terms.

For example, the two outer frequency ranges, the Low and High, sit on the opposite sides of the frequency spectrum and are simplified in Table 3.1.

Table 3.1	Frequency Comparisons
Low	**High**
20–200 Hz	5–20 kHz
Bass	Treble
Fat	Thin
Dark	Bright
Big	Small
Powerful	Weak
On the ground	In the air
Huge	Tiny
Black	White
Sad	Happy

As you can see from the table, the low-frequency range and the high-frequency range produce sounds that are opposite in description. Low-frequency areas may be described as big, fat, dark, and having power. High-frequency areas are commonly described as small, thin, bright, and sounding weak.

Boost or Cut

As previously stated, equalization is boosting or cutting a frequency or a range of frequencies by using an equalizer. Boosting a frequency increases the amplitude (volume) of a particular tone or pitch. Cutting a frequency subtracts amplitude from a particular tone or pitch. If a frequency is neither boosted nor cut, it is said to be "flat." In music production, a flat frequency response does not have a negative connotation, like a "flat note" or "flat performance" does. It simply means no particular frequency range is added or subtracted from the sound.

Slope

When a sound is equalized, the frequency that has been boosted or cut may be referred to as the "peak" frequency. Typically, this will be the frequency that is boosted or cut the most. Other frequencies are affected on either side of the peak. This area is known as the slope, or Q.

A graphic equalizer has a preset Q that cannot be changed, while a parametric equalizer gives the user the ability to change Q, if needed. A parametric EQ is a much more precise equalizer than most other EQs, because you can control amplitude, frequency, and Q.

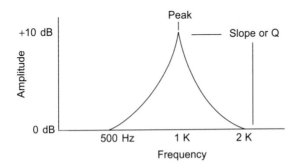

FIGURE 3.1

Low-Cut or High-Pass Filters

A button or switch often located on a console, preamp, or mic, when selected, cuts low frequencies and passes high frequencies at a predetermined setting. It does not allow you to control Q. These EQs also come in a high-cut or low-pass filter. A low cut is great to clear up any "mud" in a mix (see *muddy*, below). Try applying

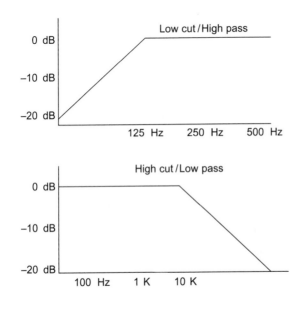

FIGURE 3.2

a *low cut* to instruments that don't have lows (electric guitar and snare drum) and a *high cut* to instruments that don't have highs (bass/kick drum). These filters can help eliminate any extraneous or unwanted frequencies in the instruments, leaving only the desired sound. Applying high and low cuts for clearing recordings of unwanted frequencies also helps in reducing the overall headroom of a track, allowing it to be louder overall without clipping (distorting).

Subtractive Equalization Is Your Friend!

Subtractive equalization is a technique used by most professional engineers to create clearer, more defined mixes. In order to have a clear mix where all instruments are heard, space will need to be made. Two sounds cannot occupy the same tone or frequency range and maintain clarity. If two sounds do occupy the same frequency range, the louder sound may *mask*, or hide, the quieter sound. Ultimately, mixing is about "crowd control." Space must be created for a sound to be heard. Many inexperienced engineers tend to add what they want to hear first. For instance, if the goal is a bigger, bassier kick drum, a novice may add more bass to the mix. A better solution is to take away from one of the other frequency areas that are dominating the sound, for example, reducing the amplitude around 600 Hz. The result will be more bass on the kick without adding destructive low-end. When mids or highs in the kick drum are cut, more bass will be present. Also, the area that has just been cut opens up more space in the mix for other instruments to be heard. This is the subtraction in subtractive equalization. This doesn't mean that frequencies should never be boosted. Start by subtracting first, and then add frequencies only as needed.

GENERAL EQ AREAS

Frequency recognition is crucial to being successful in audio production. One of the easiest ways to become familiar with the different frequency ranges and the number that goes with them is to initially divide them up in the following manner:

100 Hz – makes things bigger, fatter (kick drum).
1 kHz – adds attack, makes the sound more "In Your Face" (snare drum).
10 kHz – makes a sound airy, breathy, or brighter (hi-hat or cymbals).

❊ AUDIO CLIP 3.0

These are great EQ starting points. After you have taken out any unwanted frequencies (applied subtractive EQ'ing techniques), ask yourself, "Do I want the sound to be fatter, more up front, or brighter?" If the answer is "fatter," start at 100 Hz and adjust from there. If the answer is "more up front" or "more aggressive," boost 1 kHz. It may turn out that the correct equalization is another frequency like 2 kHz or 900 Hz. Whatever the adjustment, the key is in getting to the general area. If the answer is brighter, breathier, or airy, try boosting 10 kHz. Ultimately, a different frequency may be boosted, but adding 10 kHz should get you started.

With some generalization and through communication with the client, it will be much easier to recognize the frequency that needs to be adjusted. Locating and equalizing something quickly will hopefully keep a client happy and coming back for more!

The following are seven common EQ points of interest: subs, big/fat, muddy, boxy/hollow, in your face!, presence/clarity, and airy. Becoming familiar with these seven areas can help you locate a specific EQ point quickly. Following this section are even more terms to help you describe and communicate audio frequencies and sounds.

EQ POINTS OF INTEREST
Subs (Below 80 Hz), Low Frequencies

Frequencies below 80 Hz can make sounds huge and are referred to as "subs." Subs are often accentuated in various dance, electronic, rap, R&B, and reggae styles of music. This is the frequency area that is represented by a subwoofer. Pay close attention to this frequency area. Too much sub information can dominate a mix and make the other instruments appear weak and hidden.

AUDIO CLIP 3.1
Big/Fat (20–200 Hz), Low Frequencies

The low-frequency area generally makes sounds appear bigger and fatter. The human ear doesn't hear bass as well at lower volumes. But when we do crank it up here, terms such as big, fat, beefy, huge, and thumping are used to describe these powerful wavelengths. Too much sound here can blanket a mix, and not enough could make a mix sound weak.

AUDIO CLIP 3.2
Muddy (100–300 Hz), Low – Low-Mid Frequencies

Too much of the low and low-mid frequencies can muddy an instrument or mix. If a sound isn't very clear, or muddy, try subtracting between 100 and 300 Hz. This is especially helpful with vocals, acoustic guitars, and piano. Because close miking can cause proximity effect, a low-end boost of around 100 Hz, it is often unnecessarily present, and will likely need to be rolled off.

AUDIO CLIP 3.3
Boxy/Hollow (300–700 Hz), Low-Mid Frequencies

The frequency range 300–700 Hz is often described as boxy or hollow. This is typically an area where subtractive EQ is applied, although there are always exceptions. Kick drum mics are often designed to cut frequencies from this area. Subtracting low-mids can clean up a sound and make it more distinct, but it can also leave a sound hollow and colorless. This is not the most flattering frequency area on many instruments. An electric guitar tone, if described as boxy, has too much of this frequency range. A boxy sound can also be the result of overly *compressed* audio with a very fast attack setting, especially with a snare drum or tom.

AUDIO CLIP 3.4

In Your Face (1.5–4 kHz), Mid-Mid – Upper-Mid Frequencies

Sounds in the midrange area, especially in the mid-mid and upper midrange are best heard by the human ear. This is the area between 1.5 and 4 kHz. This also happens to be the same frequency area as a baby crying. Because we hear best in this area, sounds often appear "In Your Face." 1.5–4 kHz is often described with aggressive terms such as slap, bite, crunch, edge, and attack. Punk rock music accentuates this frequency range. Some country, folk, and acoustic music might also have more sounds in the midrange. Too much here can cause ear fatigue, whereas not enough here can make a mix or sound appear dark and distant.

AUDIO CLIP 3.5

Presence And Clarity (4–10 kHz), Upper-Mid – High Frequencies

The frequency area between 4 and 10 kHz is an area that can add presence and clarity to a mix. Often vocals are emphasized in this range to help them cut through or sit on top of a track without making the vocal sound too edgy. Note that *sibilance* is also in this area. Sibilance is associated with the "s" sound and this frequency area may need to be carefully managed with some singers. A *de-esser* is often used to help remove or soften sibilance. Inclusion of just enough information here makes a mix have presence and clarity.

AUDIO CLIP 3.6

Airy (Above 10 kHz), High frequencies

Frequencies above 10 kHz make sounds appear higher in the mix. Make sure to highlight this area to make a vocal, string, tambourine, or any other sound appear airy, breathy, thin, or bright. Transients and harmonics dominate this range. Terms associated with the sky are often used to describe this area: airy, sunny, bright, light, angelic, clouds, sparkle, and feathery. This frequency range often helps differentiate what is considered high fidelity (hi-fi) and low fidelity (lo-fi). A lo-fi recording will likely have very little, if any, frequency information above 10 kHz.

AUDIO CLIP 3.7

⚠ TIP

Pay special attention to the range of frequencies below 80 Hz. This is the most destructive frequency range and too much here can negatively affect a mix. On the other hand, just enough of this frequency range can make a mix sound huge and powerful!

ADJECTIVES: SPEAKING ABOUT AUDIO USING PLAIN ENGLISH

Additional adjectives are needed by nonengineers to describe a tone, sound, or the physical space that surrounds a sound. Although professional engineers typically use more technical descriptions, particularly in discussing frequency ranges, most engineers are familiar with interpreting a musician's request. It is likely that not all engineers will agree on the definitions used here, because of the subjective nature of describing sound, but I have full confidence that these terms, in most cases, will help you communicate and interpret ideas related to music production.

Angelic – Usually a sound buried in a large reverb and with the high-frequency range accentuated. Try applying a "cathedral" or "church" reverb and boost the extreme highs.

Beefy – Probably a sound with a lot of low and low-mid frequencies. May also be described as "thick." Guitarists often request a beefy guitar tone. When the term beefy comes up, think of a sound with a solid low-end that probably isn't too quiet in the mix.

Big – Contains a lot of low-end. Associated with the frequency range 20–200 Hz. A large room can make a big sound seem even bigger if miked from a distance. Applying certain reverbs may also make a sound appear bigger. Some musicians may also say that they want a bigger sound and all they really want you to do is turn it up!

Bite – A sound emphasized in the midrange area. If a snare is described as having bite, imagine the snare being tight and in your face. It would sit somewhere between 1 kHz and 3 kHz. Some guitar tones are often described as having bite.

Body – Depending on the frequency range of the instrument or voice, the lower frequency area would need to be dominant. Often people want to hear the body of an acoustic instrument, such as an acoustic guitar or snare drum. This request would require plenty of 100–250 Hz present in the sound.

Boomy – A sound that is boomy resides in the low and low-mid frequency range. Similar to body but is generally more of a negative term. Try cutting between 100 and 400 Hz to reduce boominess.

Brittle – As the word suggests, it means "about to break." This is seldom a flattering term. A brittle sound lacks low frequencies and highlights the upper midrange and high-frequency area above 3 kHz. Cheap digital equipment can make the high frequencies sound brittle.

Breathy – A term often associated with a vocal tone. A breathy tone would be dominated by high frequencies. Try boosting 10 kHz and up for a breathy vocal. This can be achieved by EQ and/or compression.

Chimey – Contains mostly high frequencies in the sound and would accentuate an instrument's upper harmonics. Can be found in the 10 kHz and up range. Similar to glassy.

Chunky – A chunky vocal or guitar tone would have a lot of low-mids and would likely have emphasis in the 100–300 Hz area. Similar to a thick sound.

Crispy – Think of sizzling bacon. A crispy sound would emphasize the upper-mids and highs above about 4 kHz. A crispy sound may even have some distortion present. Not usually a flattering term.

Crunchy – A crunchy sound often involves some degree of distortion or overdrive. The emphasis is in the midrange area between 1 and 4 kHz. Crunchy may be used to describe a certain guitar tone.

Deep – A sound that has a lot of depth to it from front to back, or enhanced low frequencies under 250 Hz. An example would be a deep bass tone.

Dirty – The opposite of a clean, clear sound. A dirty tone would have some amount of distortion, noise, or overdrive in the signal. Similar to fuzzy.

Distant – If a sound lacks midrange and high frequencies, it will appear further back in the sound field. Add upper-mids or high frequencies to make a sound less distant. A distant sound could also mean that it is too low in the mix or has way too much reverb.

Dry – A sound with little or no FX can be described as dry. A dry sound would not have reverb or other obvious effects present. A dry sound is most common with folk, bluegrass, and acoustic styles of music.

Dull – A sound can appear dull if it is lacking energy, highs, or is overly compressed. Add upper-mids or highs to a dull sound, or slow the attack setting on a compressor to make a sound less dull.

Edgy – Describes a sound that accentuates where we hear best, in the 1–4 kHz range. An edgy sound can make the listener feel uncomfortable like nails scratching on a chalkboard. Definitely in your face!

Fat – A fat sound accentuates the lower frequency range. A fat guitar tone, a fat vocal, a fat kick, and a fat snare sound are common requests. The fat frequency range would be around 20–250 Hz.

Fuzzy – Describes a tone that is not clear and likely has a substantial amount of overdrive or distortion associated with it.

Glassy – A glassy sound is a very thin sound with lots of apparent highs. Definitely not bassy! A clean, electric guitar tone that is extremely bright could be described as glassy.

Hard – A hard sound has a lot of midrange and accentuates the attack part of a sound's envelope. Harder frequencies are found between approximately 1 and 4 kHz.

Hollow – A hollow sound lacks a portion of its frequency range. This can be caused by phase cancellations due to room acoustics or other variances.

Hot – A sound described as hot may mean that it is turned up too loud, or the high frequency range is more noticeable. Try turning the amplitude down or rolling off some higher frequencies.

Huge – Describes a sound with excessive lows or one that is recorded in a big space.

Loose – A loose sound would lack the harder mid-mid frequency area. Loose could also describe a space or environment that has very little treatment and results in a less focused sound.

Mellow – A sound lacking upper-mids and highs is often described as mellow. A mellow guitar tone would be a darker, tubey sound as opposed to a distorted, in your face tone with a lot of 2 kHz present. Also, reverb can mellow a harder sound.

Muffled – A muffled sound would be dominated by low and low-mid frequencies in the 100–250 Hz range, resulting in a tone with less presence and clarity. Imagine singing with a blanket over your head.

Nasally – Often used to describe a vocal tone. Try cutting between 500 Hz and 3 kHz. People may also describe this same area as telephone-like, honky, or tinny.

Ringy – A ringy tone will be dominated by the mid frequencies. Snare drums are often described as ringy. A ringy tone is produced when the mic is placed close to the drum rim and both heads are tuned extremely tight. Taking away frequencies between 900 Hz and 3 kHz will likely reduce a ringy tone.

Shimmering – A sound dominated by extreme highs. A shimmering sound is in the 10 kHz and up range. To create a shimmering sounds boost the upper highs.

Shiny – Similar to shimmering. A shiny sound has plenty of highs.

Sizzly – Rarely a flattering term, sizzly describes a tone with a great deal of treble. Something referred to as sizzly can also be called glassy or crispy.

Slap(py) – Usually associated with the neck of a guitar or bass, or the kick pedal striking the head of a drum. More slap would be in the 500 Hz–3 kHz range. It can also describe a sound reflecting back, as in a slap echo.

Small – A small sound would either be overly compressed or a sound with little low or low-mid frequencies. It is likely that a small sound wouldn't have frequencies below 200 Hz. Close miking produces a smaller sound versus room miking. A snare or guitar amp may appear smaller when mic is extremely close.

Smooth – A smooth tone generally has a flatter frequency response. No frequency range would be emphasized over another. It can also be described as easy on the ears.

Soft – A soft tone typically lacks the harder midrange frequencies. Therefore, it is safe to say that extreme lows, extreme highs, or a combination, creates a softer sound. It could also refer to volume. If it is too soft, turn it up. If it's not soft enough, turn it down.

Thick – See beefy. A sound that is thick has plenty of lows and low-mids. The thick area is between 20 and 300 Hz.

Thin – A sound that is not very fat or deep. A thin sound is dominated by upper-mids and high frequencies above 4 kHz.

Tight – Tight sounds have very little reverb or environment in the sound. Close miking an instrument or voice will result in a tight sound. A tight sound is dominated by the direct signal instead of the early reflections or reverberant field. Any frequency range can be considered tight, but it is often used to describe a bass or kick drum sound that is too boomy or resonant.

Tinny – A tinny sound is a thin sound dominated by the mid-mid and upper midrange. If the vocals are described as tinny, it is not a compliment. Try cutting between 2 and 7 kHz or adding some low or low-mid frequencies.

Tiny – A sound with extreme highs and almost no lows will likely sound tiny. Not enough volume may also make a sound tiny.

Tubby – An unflattering term that describes too much low or low-mids in a sound. Try cutting between 100 and 400 Hz.

Warm – A warm tone accentuates the low and low-mid frequency range. Analog tape and tube amps are often described as warm. The opposite of a warm sound would be a cold or brittle sound.

Wet – A wet sound or wet mix would have an obvious amount of FX present. The opposite of a wet sound is a dry sound. If the vocals are drenched in reverb and the guitar sounds like it is floating in space, then you have achieved this adjective.

Here are some more helpful terms when communicating with others about the quality of sound:

If a sound lacks highs, it may be described as dark, distant, or dull.

If a sound lacks midrange, it may be described as mellow, soft, or unclear.

If a sound lacks lows, it may be described as thin, small, or bright.

If a sound has too little reverb, it may be described as dry, dead, flat, or lifeless.

If a sound has too much reverb, it may be described as wet, muddy, washy, distant, or cavernous.

If something is too loud in a mix, it may be described as in your face, up front, on top, forward, masking, dominating, hot, or separate.

If something is too quiet in a mix, it may be described as buried, masked, hidden, lost, in the background or distant.

People communicate differently when referring to the quality of sound. By learning to describe sounds in a descriptive manner, you will be able to identify and execute a sound change much more quickly than randomly searching for an unknown frequency or sound. These terms offer a starting point when equalizing, applying reverb, or executing other audio engineering functions. Without this starting point, much time will be wasted turning knobs without direction.

CHAPTER 4
People Skills. Recording Isn't All Technical!

NO KNOBS OR FADERS REQUIRED

Before we go much further, let's talk about something that is not technical, is often overlooked, and is usually underrated: people skills.

It will be helpful to you as a recording engineer if people like you. You will work with all types of people and personalities and it will benefit you if you are somewhat likeable. When an artist shares their songs with you, they are putting their very closest emotions out to be judged. A good engineer will be sensitive to the vulnerability that this can create. One of the easiest ways to influence a recording does not require a knob or fader, or any technical skills at all. It requires you to be positive and supportive of the artist(s). Not to mention, a bad attitude can negatively affect a recording session or performance.

Other non-technical skills such as time management and organization can come in handy. In addition, showing up on time and keeping track of your own schedule will be necessary skills for your success. After all, if you decide to become an audio engineer for a living, much of your competition will be musicians who aren't known for being punctual! Show up on time, or better yet, show up early, and you will only increase your odds for success.

WORD OF MOUTH

You are likely to get hired in the recording field through word of mouth. If the word is that you are difficult or unpleasant to work with, then you may not get a lot returning clients or references for new clients. If an artist has a positive experience working with you, they are likely to share their success story with other musicians and return to you the next time they want to record something. Often musicians play in more than one band or quit and move on to other musical projects. A friend of mine, John, who owns his studio and is an engineer says that he attributes his first three years of clients all to the first band he recorded. The band realized that they didn't really like each other enough to continue their initial project. Members of the original band went on to create several other bands, which chose to record with John, because they had such a good experience with him the first time around.

I have clients whom I have continued to work with over the years and clients who moved on to record with other people for a variety of reasons: they wanted to record in a new studio, they were looking for a new vibe, they were won over by studio hype, or they needed some new inspiration. When a client moves on, whatever the reason, you can't take it personally. Many of those clients will return at some point. Again, if they initially had a good experience working with you, they are likely to suggest working with you again when it is time to record their next project or recommend you to a friend's band. This is also why it is important to get to know your local music scene. Your local music scene is where you will find most of your future clients and contacts.

▲ TIP

You never want musicians to guess what you are thinking, especially if it is negative. Look and act positive.

While recording AVOID

- Staring blankly at the artist(s).
- Looking disinterested in the project.
- Saying negative things.
- Body language that says, "I would rather be anywhere but here."
- Getting into arguments.
- Picking sides if an argument or disagreement breaks out between band members.

FIGURE 4.1

FIGURE 4.2

VIBE

One of the intangible qualities that occur during a recording session is the "vibe" of the recording. Vibe is that underlying feeling that exists with the music and the people surrounding it. It is often up to the engineer to keep a good vibe going during a session by saying positive things, encouraging people to go for it, and maybe even smiling more than usual. Some musicians will light candles, put down a special carpet, dim the studio lights, or do a variety of other things to enhance the mood for a particular performance. An artist needs to be in the moment to perform the song at his or her best level, and creating a good atmosphere assists with that. If you can capture this higher level, or good vibe, it will make the recording that much better.

Pointing out things you like during a recording session contributes to good energy being captured. Don't focus on the negative. Always tell an artist what they are doing well if they ask. For instance, "that vocal take had great energy, but I think you can do it even better" instead of, "you were kind of pitchy, try again." Keep the mood light and accentuate the positive!

When designing a new studio, the designer often keeps the idea of vibe in mind. Elements that can assist in creating a good vibe are not limited to the recording space. A comfortable spot in the studio or outside that welcomes the musicians to relax, hangout, and get out of the studio can go a long way. A comfy couch, an espresso machine, a box of percussion toys, vintage equipment, or anything that says "Come and hang out here, and create music," can enhance the vibe of a session. A studio should be a place to escape from the world and create music. As an engineer, don't bring your personal problems or issues into a session. You never want to risk bringing the vibe down on a session. Besides, it isn't about you, it is about their music.

The vibe can influence how the music comes across. After all, you are recording vibrations. Are they good or bad? This is something that can't be controlled with a knob or fader or the latest technology. A good vibe always creates a better recording. As in life, if you stay positive and move forward, good things happen. This also holds true when recording music.

WHAT ABOUT MY OPINION?

There are two areas where opinions can be offered: the technical and the creative. You shouldn't be afraid to express your opinion on technical matters, such as if the guitar player is asking "Which guitar amp sounds better, the VOX or the Marshall?" Equipment choices, tone selection, and technical aspects of recording are typical areas where an engineer voices an opinion. As an engineer, your primary job is to record the music as best you can and not to put your personal preferences ahead of the artist or band. Some bands may demand or expect your opinions about the creative areas. If asked, choose your words wisely. It is possible that the honest opinion the band is asking for isn't the honest opinion the band wants to hear! Don't forget that you work for the band and not the other way around.

It really isn't your place to give your opinion on things like song selection, song structure, band politics, or picking good takes. Those are the responsibilities of the band and the *producer*. However, many recording sessions do not have a producer so you will get stuck with some of those duties. If you are asked your opinion, you certainly want to remain positive and at the minimum remain neutral. Always be encouraging. If you do have an opinion on something you really feel passionate about, present a good argument and don't get emotional. It helps to know the skill level of the artist or band you are recording, so you can accurately give them feedback. Don't get caught telling the band or artist that they could do a better take unless you actually know what that band or artist's "good take" sounds like. I made that mistake early on in my career telling a band "you can do that better" and they came back saying, "Really, that was the best we've ever performed that song." My opinion ended up crushing the band's momentum. A good take for one artist may not be an acceptable take to another.

THE INSECURE ARTIST

Artists can often feel insecure. Can you blame them? They are sharing something very close and personal with you and a potential audience. This is especially true of singers and actual songwriters, if they are performing. Put yourself in the artist's shoes: imagine singing and feeling like everyone is judging your choice of words and melody. For some artists this is a nightmare, especially for the inexperienced! If you aren't engaged with the session, and you are texting your buddies, offering a big yawn, or a glazed stare, at the end of a take, you run the risk of affecting everything from the artist's delivery to their attitude throughout the session. They may think you are in the control room bored with their performance. You will never create a good vibe this way. Even if you aren't fully into the music, the band is still owed your full attention and support. This isn't to say

Library, Nova Scotia Community College

People Skills. Recording Isn't All Technical! **CHAPTER 4** 41

that all artists need special attention. You will learn over time to recognize the clients who do need special attention and provide it.

Some vocalists perform better with a small audience in the studio: friends, family, band members, and fans. One band I recorded had between twenty and thirty people hanging around for most of the performance. It pumped them up and inspired their recording. I was willing to go along because it did positively affect the band's performance. They fed off the energy of friends and fans plus it gave their music the party beach vibe they wanted to share with listeners. On the opposite end of the spectrum, some singers appreciate a low-key environment, preferring only to work with the engineer or producer in the studio. You may have to ask guests or even band members to leave the studio while a singer performs so that the singer doesn't feel intimidated or too nervous. Feel it out. If the singer seems affected by other people listening and watching, politely ask the offenders to find something to do outside the studio. In general, an insecure artist will require more positive reinforcement than a more confident artist.

PATIENCE

A high level of patience will be required from you, if you decide to become an engineer for your career or even if you are just helping record your friends. You may have to listen to a song over and over again (sometimes a song you don't want to hear over and over again), or you may have to listen to fifty takes of a vocal that makes you want to scratch out your eyes, but you will need to take a deep breath and push forward. Exhibiting your impatience may make the artist feel even more pressure to finish up, which may make the performance worse. I once listened to a drummer attempt a drum fill for three hours! Being a drummer, I wanted to kick the person off their drum throne and do it in one take, but I didn't, and the drummer eventually got it. More inexperienced players will generally require more of your patience as they are learning how to control their instruments or voices.

FIGURE 4.3

Of course, there are just some people that require additional patience to deal with in general. Patience is a great life lesson and invaluable skill and will help you tremendously in studio situations.

COMMUNICATION SKILLS

Communication skills are key to being a successful recording engineer. You need to be able to clearly communicate a variety of things, ranging from listening to and understanding what the artist(s) or band is trying to achieve in terms of a "sound," to understanding when to *punch in* on a particular track to how to coax the best performance out of the band.

Since communication skills often involve verbal cues, it will be helpful to have a vocabulary that musicians and other engineers understand. Chapter 1 provides you with some of the necessary technical vocabulary, whereas the previous chapter provides you with descriptive terms often used in audio engineering.

To communicate well, you will have to be a good listener. You will also have to choose your words wisely. Gaining the techniques and language to deal with clients will come with time. Many artists, and people in general, aren't the best communicators. It may be up to you to get an artist to express what they are thinking to keep the session moving forward and on the right track. The only thing worse than too much feedback on a session is when an artist that says little or nothing. You may have to be more animated and extraverted than normal if your client isn't providing some necessary feedback for you to better perform your job. Maybe the artist is intimidated by the process, or by you, or just doesn't know how to express their opinion. Encourage them to speak up. Use tact, be patient, and don't be a jerk when you are trying to communicate during a session.

SONG STRUCTURE

Recognizing song structure is a common area where communication skills are necessary. It can be a tool to help effectively communicate about a client's music. Being able to recognize song structure will not only give the band confidence in you, it will help you locate *punch* in points faster. If the guitar player wants to overdub a solo and says "take me to the solo section, after the second chorus," it would benefit you to recognize and quickly locate that section of the song. Don't be afraid to take notes. Write down the time next to each section. Most recording software gives you the option to lay down markers in a song and label each part accordingly. You should do this to identify the different sections. Songs are typically made up of three distinct sections: verses, choruses, and a bridge or solo section.

A verse is usually at the beginning of the song. It generally isn't the hook but more of a section to set up the hook or chorus. The lyrics of most verses aren't repeated in a song as they are in a chorus. This may also be referred to as the "A" section.

Some songs may start off in the chorus and typically end on a chorus. Choruses are often bigger in spirit and lift the song to a higher point. From a dynamic standpoint, choruses typically are the loudest, fullest sections. This may also be referred to as the "B" section.

Some songs have bridges. Bridges are usually not repeated in a song and occur once. A bridge often occurs between two choruses but can occur almost any place in a song. A bridge is used to connect two sections of a song. This may also be referred to as the "C" section.

▲ TIP

Quickly memorizing song structure is an effective tool to help communicate during a session. If this isn't one of your strengths, you should practice the following exercise until you can easily memorize song structure.

Listen to a song and quickly identify each section:

> What is a song structure? A classic pop song will be verse, chorus, verse, chorus, bridge, and chorus.
> How many verses are there?
> How about choruses?
> Is there a bridge, and if so, when does it appear in the song?
> Is there an instrumental solo section, and if so, when does it occur in the song?
> Are there any other sections that are not verses, choruses, solos, or considered a bridge? Maybe there are pre-choruses, or short sections of only instruments before the actual verse begins. After a few listens, you should be able to identify the song structure. A good engineer can listen to a song once and will find the song structure.

Since a recording session usually involves recording songs, you will need to understand basic song structure to be able to effectively communicate with musicians. Austin Community College as well as other colleges and universities offer classes in song writing to learn these basics.

BODY LANGUAGE

Body language is the non-verbal way in which we communicate. As an engineer, be aware that musicians are interpreting your body language, whether positive or negative. Make a conscious effort not to give off pessimistic or indifferent body language, such as rolling your eyes, avoiding eye contact after a take, or shaking your head in disgust, as this can heavily influence a studio performance. Though you may be mulling over your bad choice of microphone, the musicians who just finished recording a take do not know this, and may interpret these negative cues as a reaction to their performance. Musicians should never have to worry about you being on their side. Don't give them an opportunity to create scenarios that don't exist.

You need to be not only aware of your own body language, but also cognizant of the body language demonstrated by the musicians you are recording. Being able to read this non-verbal language is an important skill to develop if it doesn't come naturally. Recognizing gestures, eye movements, and postures may help you interpret an artist's true feelings. For instance, a singer with an averted gaze may indicate he or she is unhappy with a vocal take. If you are focused and pick up on this cue, you can positively encourage another vocal attempt. This may take pressure off a musician who is too intimidated to make the request.

🔺 TIP

Specific body language to watch for:

Slumped shoulders may indicate that the musician is unhappy or disappointed in their performance or the way the session is going.

If a musician who normally has no problem making eye contact suddenly stops making eye contact, take notice. This may indicate that the musician is dissatisfied and is trying to avoid an argument or potential disagreement.

If the musicians are eyeing one another and not you while listening back to the recording, they may not be happy with how things sound.

Clearly, if the musicians are smiling, upbeat, and making eye contact with you during the recording process, they are likely happy with the vibe of the session.

SCHEDULING, TIME MANAGEMENT, AND ORGANIZATIONAL SKILLS

Good scheduling, time management, and organizational skills can help a session run smoothly.

Below is a typical schedule for four-piece band recording a few songs at my studio.

Although every session is different, this should give you a general idea of the time frame of a recording session, which in my case is usually about 12 hours. In most recording sessions, more experienced musicians get more done in a shorter period.

1. 10 A.M. I set up for the session. The drummer arrives.
2. 10:30 A.M. Set up drums, mic drums, and get sounds.
3. 11:30 A.M. The bass player arrives with the drummer, so we take advantage of it and get bass sounds too.
4. 12 P.M. The rest of the band shows up.
5. By 1 P.M. we are tracking bass and drums, listening back, and discussing how things sound.
6. 3 P.M. We finish recording the drums and bass (rhythm tracks) on the three songs the band planned to record.
   a. Everyone takes a quick break, the drummer breaks down the drums, I re-patch, and then rest my ears and brain.
7. 3:30 P.M. We start rhythm and any other guitar tracks (overdubs).
8. 5 P.M. Pizza is ordered, so we can keep working.

9. 5:15 P.M. We start recording lead vocals.
10. **Between 6 and 7** P.M. I grab a slice of pizza and I eat during vocal takes.
11. **7** P.M. We are done with recording lead vocals. We spend the remaining few hours doing a few backup vocals, percussion overdubs, and some bonus fun tracks.
12. **9** P.M. The band is tearing down and loading out. I take another quick ear break and then I get some rough mixes going.
13. Around **10** P.M. we listen back to what we did, talk it up, and celebrate our accomplishments.
14. **11** P.M. We schedule another day for mixing and I see the band off.

Recording music is like preparing a meal – it's very much about timing and organization. Just as a good cook must organize and prepare to have all dishes on the table at one time, so must a recording engineer prepare and organize to produce a finished project. Being prepared for what comes next keeps the session and the creativity flowing. Musicians aren't thrilled when they are ready to go and you are not. Mild tantrums could ensue and spoil the session's good vibe. Remember that you work for the artist as an engineer, and be ready to record.

It is unlikely that your audio engineering gig will be a Monday-through-Friday 9–5 P.M. situation. This is why you want to make sure you keep some type of work calendar. As an independent recording engineer, you will be in charge of your work schedule. No one is going to remind you that you work Friday night at the club and Saturday morning in the studio. Keeping a work calendar will ensure that you don't double book yourself or forget to show up for a gig.

Knowing what comes next is another essential element that isn't often discussed but can make a real difference in how a recording session transpires. The more sessions you run, the better you will become at keeping a session moving forward. Since most bands have a limited budget, being efficient can benefit you as an engineer. The band will be happy that they didn't go over budget and they are likely to feel more satisfied with the outcome of the session. The ways to be efficient are to plan ahead, be organized, and communicate effectively. Chapter 9 covers studio session procedures in detail.

HOW TO MEET PEOPLE TO RECORD

Another benefit of having decent communication skills and being a likeable person is that it will help you make connections and meet people to record. A great way to meet bands and musicians is to run live sound. This will be discussed further in Chapter 13. Running live sound gives you access to potential clients every night. If you aren't running live sound, go out and see local shows and get to know some people and musicians in your town. If people get to know you and like your vibe, they may trust you to record their music. As mentioned earlier, this is a word-of-mouth business. You have to get out there so that people know you and will want to record their music with you. It will be hard for people to discover you and your audio skills if you never leave your bedroom!

Playing in a band can also help you meet other musicians and bands. I met many bands that I ended up recording this way. Your first clients, or guinea pigs, may be your own band if you play in one or maybe a friend's band. Offer to record your own band or a band that you know for free. Get the experience and don't worry about the money. Your band mates or your friend's band may be a little more understanding of your limited skills and would love a free demo while you learn the ropes. Don't be afraid to tell people what you do but don't act too anxious, cocky, or pushy about it.

You may be lacking in technical skills when you first get started, so having these other skills can compensate for your lack of technical prowess. Don't underrate people skills! Good communication skills, time management, knowing your local music scene, and patience can only add to your value. Remember that recording is more than turning knobs. I can't stress how important it is to make a recording session run as smoothly as possible. Don't forget your reputation is built by word-of-mouth from your clients and your job is to serve the client. If even just one client has a problem with you or your attitude, that could affect your future business. Bite your tongue if necessary and don't take things too personally. My former band manager gave me some good advice early on in my career, "don't burn bridges, no matter what."

CHAPTER 5

Microphone Guide and Their Uses. Hey, Is This Thing On?

MICROPHONE BASICS

Microphones, or mics, are used to capture a sound much like our ears. Microphones are one of an audio engineer's finest tools. If you were an artist, microphones would be analogous to your color palette. Every microphone choice is like a stroke of the brush adding texture, tone, and color. Which microphones you choose can influence whether a recording is bright or dark, edgy or mellow, or muddy or clear. Three steps are involved in recording: capturing the sound, storing the sound, and listening back to the sound. The microphone represents the first step, capturing the sound. Getting to know how a microphone captures a particular instrument or sound takes time and experience. So what differentiates one microphone from another? Besides cost and esthetics, there are many other factors.

When choosing a microphone, there are three major categories to consider:

- Transducer/element type
- Directional characteristic
- Frequency response

What is a transducer? A transducer converts one form of energy into another. Speakers, our ears, and microphones are all transducers. A speaker converts electrical energy into acoustic energy. Our ears convert acoustic energy into mechanical energy and then finally into electrical energy, which is sent to our brains. A mic converts acoustic energy into electrical energy.

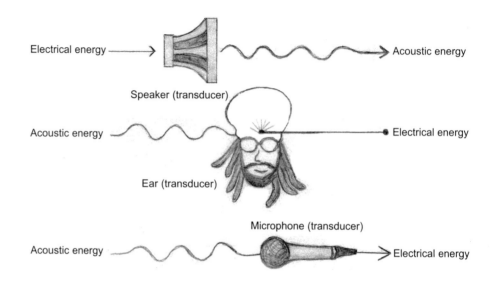

FIGURE 5.1

Transducers are often considered the weakest link in the recording chain. This is because they exist in almost every stage of the signal path. For instance, there are quite a few transducers involved in recording an electric guitar plugged into an amplifier. First, the pickup in the electric guitar is a transducer. This transducer takes the acoustic vibrations from the guitar strings and converts them into the electrical equivalent. This electrical signal is sent to the amplifier and converted back to acoustic energy through the amp's speaker. The mic placed in front of the amplifier converts the acoustic energy from the speaker back into electrical energy. The signal is then sent to headphones or monitor speakers and converted back to acoustic energy. Finally, acoustic energy travels to our ears where it is converted back to electrical energy, and then sent to our brains to be processed as sound. In this particular scenario, we have identified five transducers in the signal path of an electric guitar recording: the guitar pickup, the amp's speaker, the mic placed in front of the amp, headphones/studio monitors, and our ears. If any one of these transducers is flawed or inadequate, the end result of capturing a quality guitar sound could be jeopardized.

Basic Vocabulary

Transient – A short, quick burst of energy that is non-repeating. Commonly associated with the attack of percussive instruments. However, all music contains transients. In speech, transients are associated with consonants. Transients are typically weak in energy and associated with the higher frequency range.

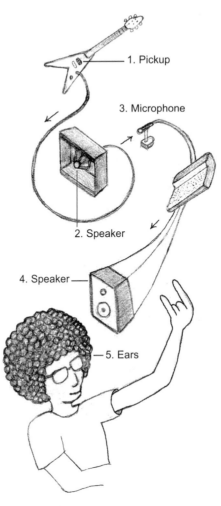

1. Pickup

3. Microphone

2. Speaker

4. Speaker

5. Ears

FIGURE 5.2

Transient response – How quickly the microphone reacts to a sound wave and specifically to those transients just described. This differentiates one mic sound from another. Condenser mics typically have the best transient response.

Preamp – A control used to boost an audio signal's level. Microphones are plugged into preamps. The preamp knob is turned up to a useable and desired signal level. This level is generally sent to a recorder or a set of speakers (Figure 5.3).

Leakage aka bleed-over – The amount of sound that bleeds into the source being recorded. Leakage could be anything from room ambience to another instrument sound. This is common when a full-band performs together with all the instruments in the same room or stage. Microphones with tighter pickup patterns provide less leakage from other sources and the environment. A mic with a hypercardioid pickup pattern provides the best isolation and prevents the most bleed-over from other sounds.

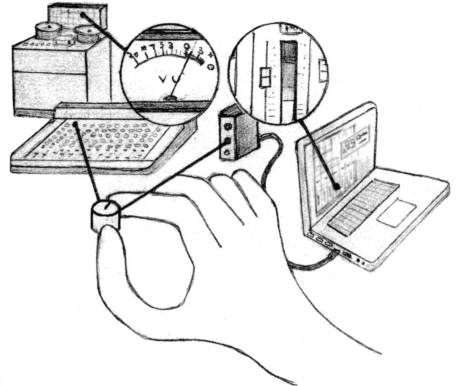

FIGURE 5.3

Pop filter – A nylon screen around a hoop or a perforated metal disk placed in front of the microphone in order to avoid plosive "b," "p," and "t" sounds that create destructive air movement. Pop filters are usually mounted on the mic stand with the mic or on a separate mic stand and are placed a few inches away from the mic. Pop filters are typically used when miking vocals. Some mics have built-in screens but an external filter is still needed. Filters also keep saliva off the actual microphone.

Common switches found on microphones:

dB pad – Used to attenuate gain on a mic. Typically found on condenser mics. Usually specifies the amount of gain cut in dB: −10, −15, −20. Use this when miking louder sounds. Many mic preamps also have this function. This pad can identify a condenser mic, although not all condensers mics have dB pads.

Low-cut or high-pass – Used to roll-off low frequencies and pass highs. Typically found on mics with cardioid pickup patterns. The user can select where the cut takes place, such as 20, 75, and 100 Hz. Great for non-bass instruments. Can help clear up a sound, cut out mud, and reduce low frequencies.

Pickup pattern selector – Allows the user to choose the directional characteristic of the microphone.

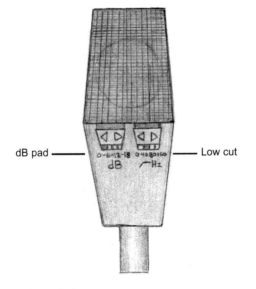

dB pad ——————— ——————— Low cut

FIGURE 5.4

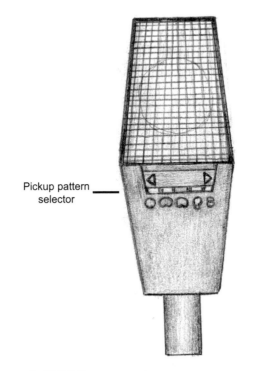

Pickup pattern _____
selector

FIGURE 5.5

Shock mounts – Helps isolate the mic from traffic, thumps, and microphone stand transmission of sound. Included with many microphones, especially condenser mics.

Transducer Types

The first category to consider when choosing a microphone is the transducer type.

Dynamic mics are built tough and can handle loud sound pressure levels. General characteristics and uses of a dynamic mic:

- Very rugged, which makes it the main microphone type utilized in live sound.
- Great for loud things such as amplifiers, kick and snare drums, toms, and horns.
- Generally used in conjunction with close miking. This makes sense, considering that they are extremely rugged (they may be hit by a drumstick or two) and good with high sound pressure levels (a cranked amplifier).
- General frequency response is between 40 Hz and 15 kHz.
- Common dynamic microphones: Shure SM57 or SM58, AKG D 112, Sennheiser MD 421, Electrovoice RE 20, Audix D1 – D6.

Condenser microphones are best at reproducing transients and generally provide greater clarity. General characteristics and uses of a condenser mic:

- Excellent transient response; therefore, great for reproducing higher frequencies and quiet sounds. Adds "air" to a sound. Because the transducer is lighter, it reacts better to weaker sounds (highs).
- Most require external power known as phantom power. Phantom power is usually located near the microphone preamp. It is often labeled +48 V. Engaging this button will supply a condenser microphone with its required charge. Some condenser mics can use 9 V or AA batteries to supply the charge.
- Fragile. Condenser mics, unlike dynamic mics, are considered fairly fragile. Hitting a condenser mic with a drumstick or dropping it could be the end of this mic's life.
- Often have a dB pad. A db pad, used to attenuate input into the microphone, can help identify a mic as a condenser microphone.
- Available as a Large-Diaphragm Condenser (LDC) or Small-Diaphragm Condenser (SDC).
- Small-diaphragm condensers (SDC) are best at reproducing transients. Great for recording acoustic instruments, high hat, overheads on drums, room mics, flute, and shaker. Sometimes referred to as pencil condensers. Common SDC microphones: AKG C 451 B, Shure SM81, Rode NT5, Neumann KM184, and MXL 600.
- Large-diaphragm condensers (LDC) are typically my choice for recording vocals. They add presence to vocals and usually help vocals sit in the right place. A LDC also exhibits better response to low frequencies and can help to fatten up an instrument or voice. It should be mentioned that ribbon mics and dynamic mics can also record vocals quite effectively.

- General frequency response – 20 Hz–20 kHz.
- Common LDC microphones: AKG C414; Neumann U 47 or U 67; AT4050; and Shure KSM27, 32, or 44.

Ribbon microphones are often used to take the "edge" off an instrument's tone. General characteristics and uses of a ribbon mic:

- Extremely fragile.
- Great for digital recording!
- Usually darkens and makes a sound appear smoother.
- Great for room sounds, acoustic instruments, drum overheads.
- Not meant for outdoor application (they don't like wind) although there are models that can be used in live sound or outdoor situations.
- Can often provide a "retro" type sound.
- Typically excel in the low-mid frequency range.
- General frequency response is between 40 Hz and 15 kHz.
- Common ribbon microphones: Royer R-121 or R-122, RCA 44, Beyerdynamic M 160, and Cascade Fat Head.

⚠ TIP

Try setting up three mics, one dynamic, one condenser, and one ribbon. If you don't have all the three available to you, just use what you have. Mic a tambourine or shaker from about 1 ft away. These instruments produce a lot of transients and have very weak energy. Note the results. Which mic changed the source the most? Which one darkened the sound? Brightened the sound? Sounded most like the source? The lightest, most sensitive transducer will likely reproduce more highs. Note the results for future sessions.

Directional Characteristics

2 Pickup Patterns

The next category to consider when choosing a microphone is the directional characteristic of the microphone, also referred to as a mic's pickup or polar pattern. The directional characteristic of the mic determines the direction from which the mic will be able to pick up the sound. When the mic is positioned as designed, this is called "on axis."

The pickup pattern also determines how well the recorded sound is isolated. Determining whether the microphone is a side or top address is also important.

Cardioid pickup patterns are the most common. Their pickup pattern is heart-shaped. Cardioid mics are sometimes referred to as unidirectional. With a top address microphone the sound is rejected from the rear of the microphone. Lots of cardioid pattern mics are used with live sound. This is because the mic rejects sound from the monitor and decreases the chance of a feedback loop.

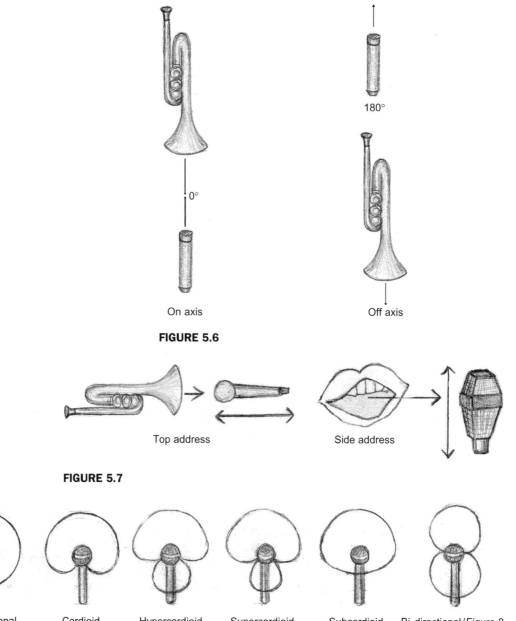

On axis

Off axis

180°

0°

FIGURE 5.6

Top address

Side address

FIGURE 5.7

Omnidirectional   Cardioid   Hypercardioid   Supercardioid   Subcardioid   Bi-directional/Figure 8

FIGURE 5.8

Supercardioid has a pattern that is tighter than a cardioid pickup pattern. It rejects more from the sides, but picks up a small amount of sound from the rear. Most beta series mics have a supercardioid pickup pattern, such as a Shure beta57.

Hypercardioid is the tightest pickup pattern of all. It provides the most isolation for an instrument and virtually excludes the environment.

Subcardioid has a pickup pattern that is a cross between omnidirectional and cardioid. It rejects some sound from the rear of the mic.

Microphones with cardioid pickup patterns exhibit proximity effect. Proximity effect is a low-end boost of 100 Hz + 6 dB when you get within a ¼" of the diaphragm. Basically, you get a bassier tone as you place the mic closer to the source.

There are three ways to avoid proximity effect.

1. Back away from the mic.
2. Use a low-cut filter or roll-off. Low-cut filters are located on the microphone itself or near the preamp section.
3. Use a microphone with an omnidirectional pickup pattern. Mics with an omnidirectional pickup pattern do not exhibit proximity effect.

▲ TIP

Take advantage of proximity. A beefier, bassier tone can be achieved by placing a cardioid pickup pattern mic right next to the source. Beware, this can also make sounds muddy and overly bassy.

Directional characteristics determine how isolated a sound will be from other sounds and how much environment will be heard in the recorded sound.

Bi-directional, or Figure 8, pattern picks up sound from the front and the back of the mic and rejects sound from the sides. Ribbon microphones tend to be bi-directional, although they come in all pickup patterns.

Omnidirectional pattern picks up sound pressure equally from all directions. It does not exhibit proximity effect and tends to have more "environment" in the sound and a flatter frequency response. It is often used when a reference mic is needed. Most lavalier mics are omnidirectional.

It is worthwhile to mention that mics can be a combination of transducer types and directional characteristics. Any transducer can have any pickup pattern. Some microphones will have a pickup pattern selector switch where you can select different patterns. These switches are found on multi-pattern mics such as the AKG C 414, AT 4050, and Cascade Elroy.

The transducer type influences the coverage area the mic picks up. If a condenser, dynamic, and ribbon mic with the same pickup patterns are placed in the same room, a condenser mic will pick up a much larger area.

Frequency Response

The last category to consider when choosing a microphone is the frequency response. Frequency response refers to how well a particular mic is able to respond to all the frequencies that strike it. Put simply, does the final result sound like the original source or does it darken or brighten the original source?

Frequency response can be divided into two categories: linear and non-linear. Since a microphone is used to reproduce the source of a sound, the concern

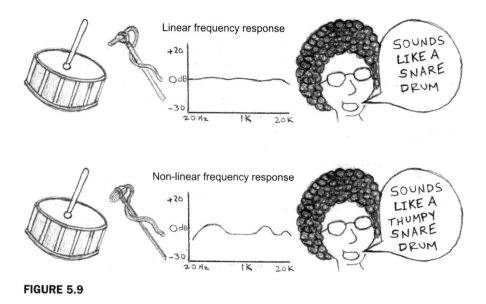

FIGURE 5.9

is how the microphone will represent and capture the original sound. A microphone with a non-linear frequency response will affect the source and alter the original sound. Will it make the original sound brighter or edgier? A microphone with a non-linear frequency response will have peaks and valleys in its frequency response. Some engineers refer to this as "coloring" the sound.

⚠ TIP

For example, you are miking a guitar amp that is very bright, with a lot of mid-highs/highs in the sound. Your intention is to darken the sound, or at least, not accentuate the higher frequencies. In these circumstances, you might choose a microphone that has a frequency response to boost the lows or reduces the higher frequencies, most likely using a dynamic or ribbon mic to help color the sound. If you had chosen a mic with a brighter frequency response, like a condenser, the result would have been an extremely bright recording. Coloring a sound can be good or bad, depending just on what you are trying to achieve.

A microphone with a linear frequency response is said to have flat frequency response. That means that it reproduces the sound in a much more transparent way. A mic with a flatter frequency response will be the best choice if the source does not require any tone alterations. Most mics come with a frequency response chart or at least specifications. You can also find additional technical information on the manufacturer's website.

A microphone with a non-linear frequency response will color or alter the way the source sounds and will not capture the source in a transparent way.

A microphone with a linear response will capture the sound in a much more transparent way altering the tone little, if any.

MICROPHONE PLACEMENT
Where to Place the Microphone?

Now that you know a little more about how microphones work, it is time to discuss where to place them.

Mic placement is an art and is one of the engineer's most important tools. Mic placement is as important as mic selection. A cheap mic placed in the right place can be as effective as an expensive mic placed in a wrong place.

Before you determine a microphone's placement, consider where the performers will be placed according to sight, feel, and acoustics. Don't overlook the importance of sight and feel, because both can heavily influence a musician's performance. Musicians use non-verbal cues to communicate when they are performing. Make sure the musicians have a line of sight between them in the studio. It is also important for the musicians to feel comfortable where they are placed. For instance, placing a musician who is claustrophobic in a small isolation booth probably will not result in a good performance, no matter what mic is used. Finally, have the musicians play their instruments in different locations and hear where they sound best. Many instruments will jump out when placed in the optimum spot in a room.

Next, before deciding mic placement, it needs to be determined where the majority of the sound is coming from and the direction it is being projected. Note: some instruments' bodies have a single area from which the sound is being projected (a trumpet), while others may have several areas from which the sound is being projected (a saxophone).

The easiest thing to do is simply listen to the instrument to find out where it sounds best and start by placing a mic there. After all, a microphone is being used to represent our ears. Be aware that moving a mic even a small distance can dramatically change the sound. It is often hard to determine how good something sounds without comparing it with another sound. Don't be afraid to move the mic around and compare it with other mic positions. Recording these positions and comparing them can be helpful.

Keep in mind that sound is divided into three successively occurring categories: direct path, early reflections, and reverberation. The direct path is the quickest path to the listener and it helps determine where the sound is coming from and provides a clear sound. Early reflections occur right after the direct path and clue us in to the surface(s) of the environment. Finally, the reverberant field is the last part of a sound heard and helps us identify the size of the space or environment. This is especially important when deciding where to place a mic. The term "reverb" is described in more detail in Chapter 7.

Four Fundamental Styles of Mic Placement According to Distance

Here are some great starting points for miking instruments. Note that each position yields a different result and each position either accentuates the sounds

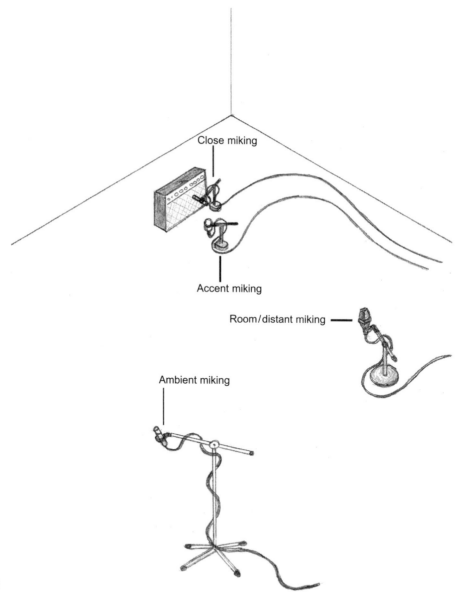

FIGURE 5.10

direct path, its early reflections, its reverberant field, or any combination of the three.

1. **Close miking** – placing the mic from 0 to 1 ft from the source. This provides the best isolation for the source and reduces the environment in the sound the most. Picks up the "direct path" primarily with little or no early reflections or reverberation present in the sound. Provides an "in your face sound" that can be described as tight and clear. Ultimately, it isn't very natural (how many times have you stuck your head in a kick drum?); that is why it is common to combine close miking with other techniques or to add artificial reverb to the sound. Watch for proximity effect when close miking.

 **AUDIO CLIP 5.0**

2. **Accent miking** – 1–3 ft. Common with live sound. As the name suggests, it helps accent a sound. Great for soloists and ensembles. This miking technique captures both the direct path and early reflections, with a small amount of reverberation in the sound.

AUDIO CLIP 5.1

3. **Room or distant miking** – 3 ft – Critical Distance is the point where the sounds direct path and reverberant field are equal. I would consider this the most natural miking technique. This is the distance we would listen to someone playing drums or playing a guitar through an amplifier. This technique is often combined with close miking. Room miking doesn't provide much isolation and will have a good amount of the environment present in the sound. When miking this way, the sound will be made up of the direct path, early reflections, and reverberant field more evenly.

AUDIO CLIP 5.2

TIP

Try miking drums with a single large diaphragm condenser from 3 to 6 ft away. Drums will sound big and likely won't interfere spatially with close mics used on other instruments. Because the drums cannot be individually balanced, a good or an interesting sounding environment is essential with this method as well as good sounding drums.

4. **Ambient miking** – Miking beyond the critical distance. Like room miking, it is often combined with close miking to retain clarity. With ambient miking the reverberant field is dominant and the source is usually less intelligible and clear.

AUDIO CLIP 5.3

Two Great Stereo Miking Techniques to Know

3 Phase

What is stereo miking? Why would I use it?

Stereo miking involves using two or more microphones to represent an image. It makes a sound wider, bigger, and thicker. Actual stereo miking provides greater localization than taking a mono mic and applying a stereo reverb/effect to the sound. In the age of digital recording, stereo miking will make an mp3 sound a lot more interesting and the sound will have more depth and character to it.

With stereo miking you will record each mic to a separate track. You can then pan each track to opposite directions or anywhere in between. Make sure you hit the MONO button on your console, audio interface, or software application to insure that the mics are in phase.

FIGURE 5.11

XY is a great technique to use when miking yourself playing an instrument. This technique is considered coincident miking, meaning that the transducers are placed in the same place. XY provides a pretty good stereo image with little or no phase issues. Because time is not an issue between the transducers, phase is not an issue. This technique translates well in mono. Typically, two like microphones are used for this technique. SDC are often utilized in this technique, such as a pair of SM81s.

Spaced pair is a common stereo miking technique involving two mics spaced apart to capture a sound. Unlike XY, spaced pair can have some phase issues. This is due to the fact that time does become an issue with the spaced pair setup. Again, any time there is a time difference between two or more microphones, you will likely have some phase issues. This technique typically provides a more extreme stereo image than XY but doesn't translate as well in mono. Make sure that the two mics are placed three times the distance from each other as they are placed from the source. This will help with phase issues. Keep in mind the 3:1 rule when using multiple microphones.

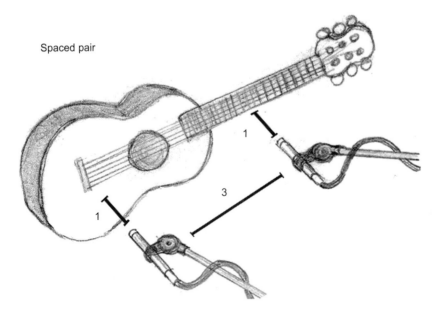

Spaced pair

FIGURE 5.12

⚠ TIP

When I am not sure where to mic an instrument, I usually ask the musician where it has been mic'd before or if they have a preference. When you are unfamiliar with an instrument, this is a good starting point. Besides that, it will also make the musician feel included in the recording.

Direct Box

In some cases, you may use a direct box instead of miking up an instrument. A direct box is most often used with bass guitar and electric keyboards. Direct boxes are also used with stringed instruments such as violin, cello, and acoustic guitar. It eliminates the need for a mic by taking the instruments line out and converting it into a mic input. This direct signal is supplied by the instrument pickup or line out. Direct boxes can be either passive or active and some require phantom power.

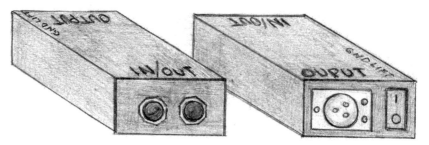

FIGURE 5.13

Quick Mic Setups

HOW TO MIC A DRUMSET

Unless you have an unlimited amount of time, keep it simple when miking a drumset. Many classic drum sounds used only a few mics. Of course, that technique won't work for all styles of music but it works for most. Dynamic mics are typically used for close miking of drums and overheads and room sounds are represented with condenser or ribbon mics. In the following illustrations, I demonstrate four ways to mic up a drumset. Try using a single large diaphragm condenser or ribbon mic and placing it about waist high 3 ft from the kick drum. Raise the mic up if you want less kick and you desire more snare drum. Close miking a kit takes time but it can result in a tight, dry sound. Place a dynamic mic a few inches away from each drumhead at an angle of 45–60 degrees. Place a stereo pair of condenser or ribbon mics in the room. Remember that if you use a spaced pair, don't forget to apply the 3:1 rule discussed earlier in this chapter. A simple way to get a quick and decent drumset sound is by placing one mic over the center of the kit to capture the snare, toms, and cymbals. Try using a condenser or ribbon mic for this. For the kick drum, try placing a dynamic mic inside the drum and adjusting it until you get the desired tone. A great stereo drum sound can be achieved easily with the top/side miking technique. With this method place two mics equal distance from the snare (a mic cable is handy to measure the distance), one mic is placed over the drummer's shoulder and the other mic is placed lower on the floor tom side. Pan the mics opposite directions. Since the two mics are equal distance from the snare, the snare will be in phase and no mic will be needed to hear the snare drum. Place a mic inside the kick drum for the low-end.

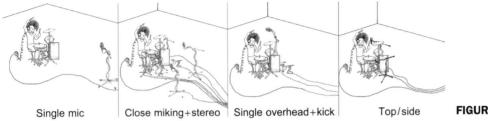

Single mic | Close miking+stereo | Single overhead+kick | Top/side **FIGURE 5.14**

HOW TO MIC A BASS

The most common practice is to use the direct sound from the bass and avoid using a mic. Try a direct box or a direct output from the bass head. If you do mic the bass amp, try a dynamic mic up close and/or a condenser mic in the room. To get tricky, you could blend all the three sounds: the direct signal, the close mic, and the room sound.

FIGURE 5.15

HOW TO MIC A GUITAR AMP

The fastest way to mic up a guitar amp is to place a close mic on the speaker. If you want a brighter, tighter sound, position the mic close to the center of the speaker. A darker, looser tone is achieved by placing the mic more to the outer edge of the speaker. If you have more than one track for guitar at your disposal, try combining the close mic with a ribbon mic that is placed 1–3 ft away from the amp. Some engineers like to record a DI signal from the guitar at the same time as the microphones. Although you may not use it in the final mix, it will be invaluable later on if you need to reamp.

FIGURE 5.16

HOW TO MIC AN ACOUSTIC GUITAR

The easiest and fastest way to get a good acoustic guitar sound is to place a mic about 5" away from the 12th fret (the double dots on a guitar).

FIGURE 5.17

HOW TO MIC A VOCAL

Every voice is unique. Once you figure out what mic sounds best on the singer, place the mic about 6" away from the singer. To determine what mic sounds best, setup two mics at a time, side by side, and compare the results. Repeat this until you are satisfied.

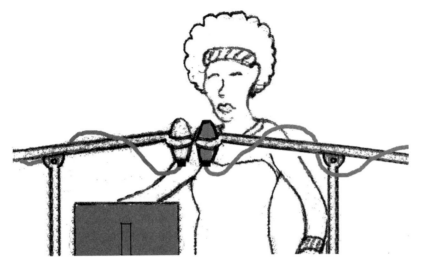

FIGURE 5.18

HOW TO MIC BACKING VOCALS

Try placing an omnidirectional mic in the room and have the musicians circle around the mic. Move the singers around until you have your desired blend. Try doubling the backup vocals and panning the two tracks for an even bigger, thicker sound.

FIGURE 5.19

For more on microphone techniques check out *Practical Recording Techniques*, Bruce Bartlett, Focal Press, 2008.

MICROPHONE GUIDE AND USES

There are more microphone choices than ever before. There could be a whole book on just microphones and their uses. The following section will feature common mics and touch on a cross section of what is out there. I have incorporated mics that a home recordist could afford as well as a fewer high-end mics that you may encounter in a professional recording studio. This is only a taste of what is out there.

The guide below provides a picture to help identify the mic, a price guide, the microphone's transducer type and pickup pattern, specifically what the mic was designed for, what instrument to try it on, and a frequency response graph.

Price key:

$ – under $200
$$ – $200–$500
$$$ – $500–$1000
$$$$ – over $1000

Dynamic Mics
AKG D 12

FIGURE 5.20

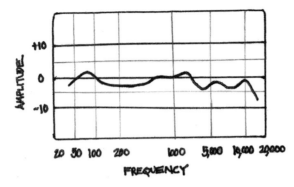

FIGURE 5.21

$$
Dynamic
Cardioid
30 Hz–15 kHz
Designed with a special "bass chamber" that gives this large-diaphragm mic great bass response. Exhibits a slight boost at 60 and 120 Hz. Introduced in 1953.
Try on: trombone, tuba, acoustic bass, and kick drum.

AKG D 112

FIGURE 5.22

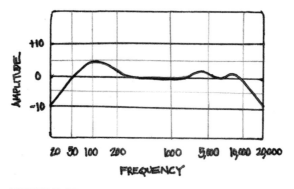

FIGURE 5.23

$/$$
Dynamic
Cardioid
20 Hz–17 kHz
Designed for kick drum.
Try on: kick drum, trombone, and bass amp.

AUDIX D6

FIGURE 5.24

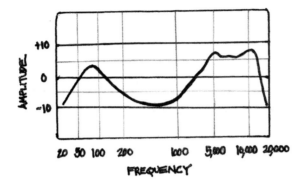

FIGURE 5.25

$
Dynamic
Cardioid
30 Hz–15 kHz
Designed for Kick drum.
Try on: kick drum, floor tom, bass amp, and Leslie speaker bottom.

BEYER M 201

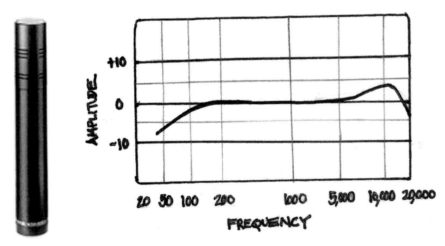

FIGURE 5.26   **FIGURE 5.27**

$$
Dynamic
Hypercardioid
40 Hz–18 kHz
Designed to work well with both live and recording sound on hi-hats, snare
drums, rack toms, and percussion.
Try on: snare drum, hi-hat, drum overhead, piano, acoustic guitar, and strings.

ELECTROVOICE RE20 VARIABLE-D

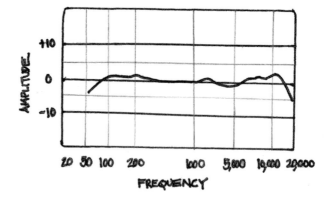

FIGURE 5.29

FIGURE 5.28

$$
Dynamic
Cardioid
45 Hz–18 kHz
Designed for radio and television broadcast announcers.
Try on: kick drum, bass amp, vocals, horns, and guitar amp.

SENNHEISER E 609

$$
Dynamic
Supercardioid
50 Hz–15 kHz
Designed for percussion and brass instruments.
Try on: guitar amps, snare drum, and live vocals.

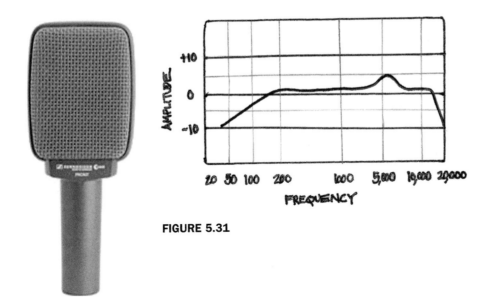

FIGURE 5.31

FIGURE 5.30

SENNHEISER MD 421

$$
Dynamic
Cardioid
30 Hz–17 kHz
Designed for most instruments, group vocals, and radio broadcast announcers. Introduced in 1960.
Try on: kick drum, snare drum, toms, and any other percussion. Often works great on bass amps, guitar amps, and horns.

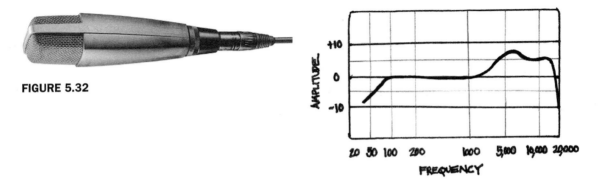

FIGURE 5.32

FIGURE 5.33

SHURE BETA 52A

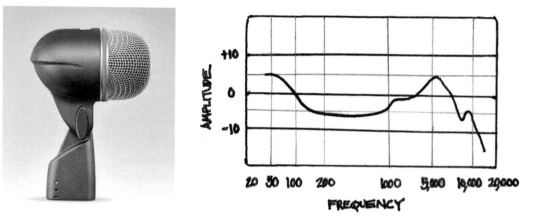

FIGURE 5.34   FIGURE 5.35

$
Dynamic
Supercardiod
20 Hz–10 kHz
Designed for miking a kick drum live.
Try on: kick drum, bass amps, and other low-end instruments.

SHURE BETA 57A

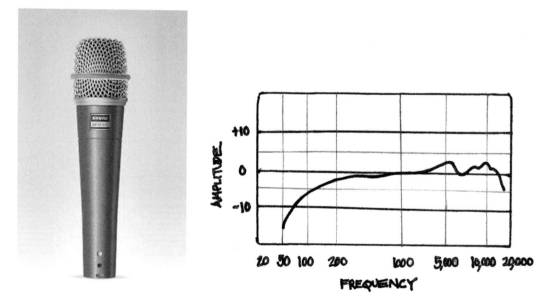

FIGURE 5.36   FIGURE 5.37

$
Dynamic
Supercardioid
50 Hz–16 kHz
Designed for drums, guitar amps, brass, woodwinds, and vocals.
Try on: snare drum (helps eliminate high-hat bleed), congas, and other percussion.

SHURE SM57

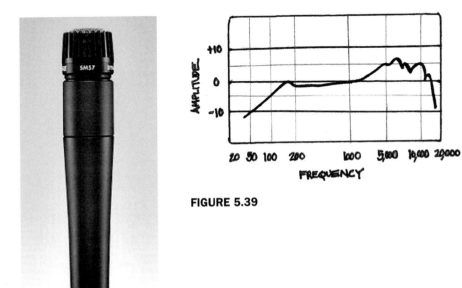

FIGURE 5.39

FIGURE 5.38

$
Dynamic
Cardioid
40 Hz–15 kHz
Designed for amps and instruments. Excels in live sound but is common in the studio; it's rugged and inexpensive.
Try on: snare drum, toms, guitar or keyboard amp, horns, and vocals.

SHURE SM58

$
Dynamic
Cardioid
50 Hz–15 kHz
Designed for vocals, especially for live performance.
Try on: vocals, backing vocals, horns, strings, and snare drum.

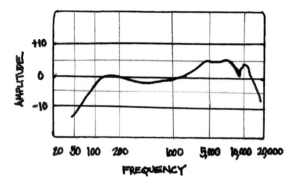

FIGURE 5.41

FIGURE 5.40

SHURE SM7B

FIGURE 5.43

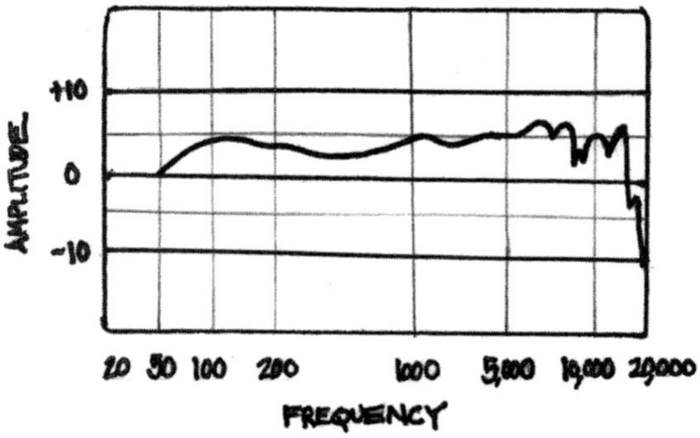

FIGURE 5.42

$$ Dynamic
Cardioid
50 Hz–20 kHz
Designed for radio broadcast announcers and vocals.
Try on: kick drum, bass amp, loud vocals, and horns.

Condenser Mics

AKG C 414

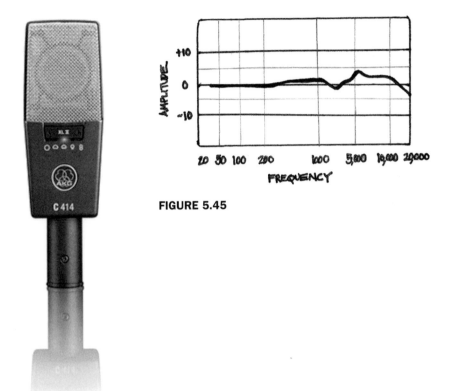

FIGURE 5.45

FIGURE 5.44

$$$
Large-Diaphragm Condenser
Omni, wide Cardioid, Cardioid, Hypercardioid, Figure 8
20 Hz–20 kHz
Designed for recording detailed acoustic instruments.
Try on: vocals, group vocals, acoustic guitar, room sounds, and drum overheads.

AKG C 12

$$$$
Large-Diaphragm Tube Condenser
9 selectable patterns from Omni to Figure 8
30 Hz–20 kHz
Designed for recording brass, strings, and vocals.
Try on: acoustic guitar, flute, strings, vocals, and guitar amp.

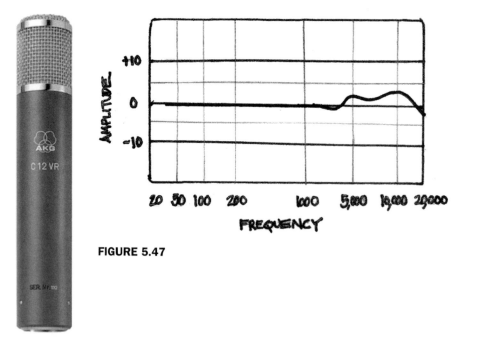

FIGURE 5.47

FIGURE 5.46

AKG C 451 B

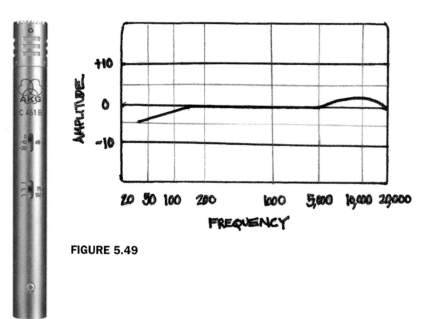

FIGURE 5.49

FIGURE 5.48

$$/$$$
Small-Diaphragm Condenser
Cardioid
20 Hz–20 kHz
Designed for capturing rich transients such as with an acoustic guitar or drum overheads.
Try on: piano, hi-hat, cymbals, acoustic guitar, and percussion.

AUDIO TECHNICA AT4050

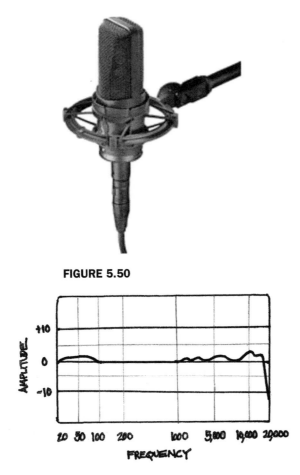

FIGURE 5.50

FIGURE 5.51

$$/$$$
Large-Diaphragm Condenser
Omni, Cardioid, Figure 8
20 Hz–18 kHz
Designed for vocals, piano, strings, and drum overheads.
Try on: group vocals, lead vocals, room sounds, and acoustic guitar.

AVANTONE CV28 TUBE MIC

FIGURE 5.52

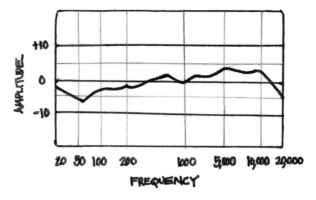

FIGURE 5.53

$$
Small Diaphragm Tube Condenser
3 Interchangeable capsules: Omni, Cardioid, Figure 8
20 Hz–20 kHz
Designed for acoustic instruments, drum overheads, percussion, and piano.
Try on: acoustic guitar, violin, mandolin, and drum overheads.

AVENSON STO-2

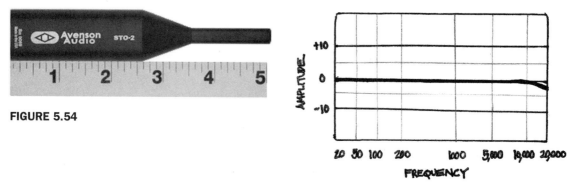

FIGURE 5.54

FIGURE 5.55

$$/$$$
Small-Diaphragm Condenser
Omni
20 Hz–20 kHz
Designed to be used as a matched pair.
Try on: room sounds, drum overheads, and acoustic guitar.

BLUE DRAGONFLY

FIGURE 5.56

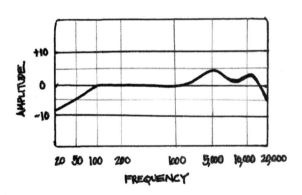

FIGURE 5.57

$$$
Large-Diaphragm Condenser
Cardioid
20 Hz–20 kHz
Designed for high-frequency sources like alto and soprano vocals, percussion,
electric guitar, and drum overheads.
Try on: vocals, strings, and room sounds.

EARTHWORKS TC30

FIGURE 5.58

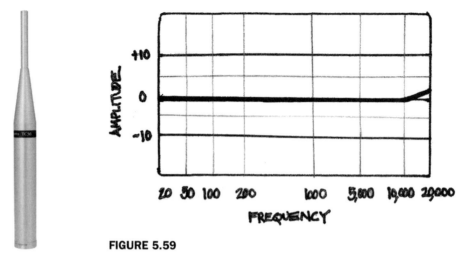

FIGURE 5.59

$$$
Small-Diaphragm Condenser
Omni
9 Hz–30 kHz
Designed for recording louder sources.
Try on: drums, amps, live club recordings, and room sounds.

MOJAVE AUDIO MA-200

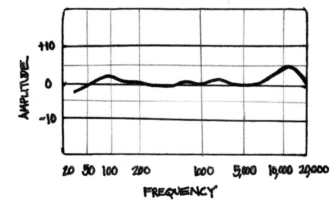

FIGURE 5.60

FIGURE 5.61

$$$/$$$$
Large-Diaphragm Vacuum Tube Condenser
Cardioid
20 Hz–20 kHz
Designed for anything you would use an expensive large-diaphragm condenser on.
Try on: vocals, voice-overs, acoustic instruments, drum overheads, and piano.

MXL 990

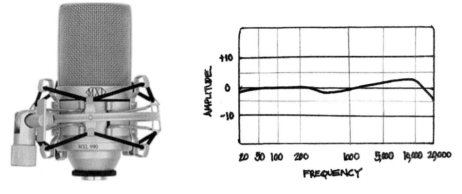

FIGURE 5.62   **FIGURE 5.63**

$
Medium-Diaphragm Condenser
Cardioid
30 Hz–20 kHz
Designed as an affordable vocal mic and for acoustic guitars.
Try on: vocals, acoustic guitars, and room sounds.

NEUMANN TLM 103

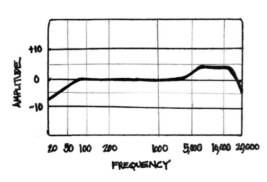

FIGURE 5.64   **FIGURE 5.65**

$$$/$$$$
Large-Diaphragm Condenser
Cardioid
20 Hz–20 kHz
Designed for quiet acoustic sources.
Try on: acoustic guitar, room sounds, flute, and vocals.

NEUMANN U 87

FIGURE 5.66

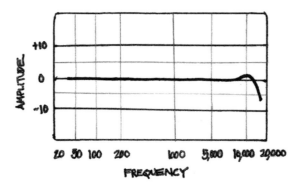

FIGURE 5.67

$$$$
Large-Diaphragm Condenser
Omni, Cardioid, Figure 8
40 Hz–16 kHz
Designed for orchestral recordings, as a spot mic for individual instruments, and vocals for music or speech.
Try on: vocals, acoustic guitar, room sounds, and drum overheads.

PELUSO MICROPHONE LAB 22 251

FIGURE 5.68   **FIGURE 5.69**

$$$$
Large-Diaphragm Vacuum Tube Condenser
9 switchable patterns from omni to bi-directional
20 Hz–24 kHz
Designed after and inspired by the vintage "251."
Try on: vocals, acoustic guitar, room sounds, and upright bass.

RODE NT5

FIGURE 5.70

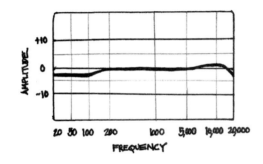

FIGURE 5.71

$$
Small-Diaphragm Condenser
20 Hz–20 kHz
Cardioid
Designed for small choirs and ensembles, cymbals, drum overheads, and acoustic instruments.
Try on: acoustic guitar, hi-hat, drum overheads, and live recordings.

SHURE KSM27

FIGURE 5.72

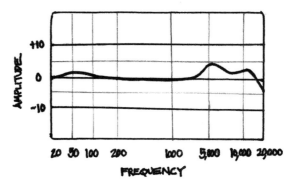

FIGURE 5.73

$$
Large-Diaphragm Condenser
Cardioid
20 Hz–20 kHz
Designed for general purpose and a wide range of applications.
Try on: a nasally vocal, acoustic guitar, flute, and room sounds.

SHURE KSM44

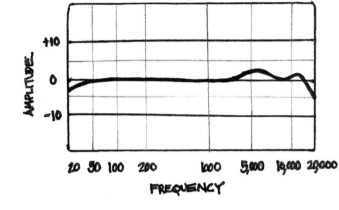

FIGURE 5.75

FIGURE 5.74

$$$
Large-Diaphragm Condenser
Cardioid, Omni, Figure 8
20 Hz–20 kHz
Designed for woodwinds, acoustic guitar, studio vocals, brass, percussion, strings, acoustic bass, piano, and voice-overs.
Try on: flute, vocals, and acoustic guitar.

WUNDER AUDIO CM7 FET

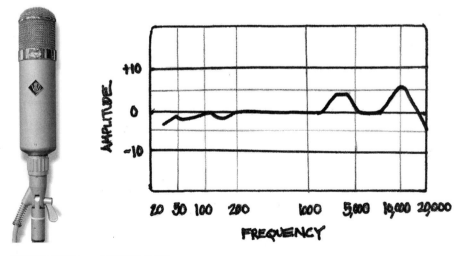

FIGURE 5.76 **FIGURE 5.77**

$$$$
Large-Diaphragm Condenser
Cardioid, Omni, Figure 8
30 Hz–20 kHz
Designed for larger-than-life kick drum and bass guitar sounds and thick vocals.
Try on: kick drum, bass amps, vocals, and for beefy room sounds.

Ribbon Mics

BEYERDYNAMIC M 160

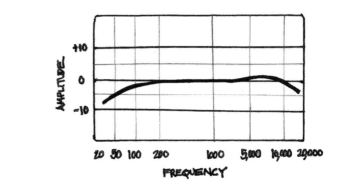

FIGURE 5.78   **FIGURE 5.79**

$$/$$$
Double ribbon
Hypercardioid
40 Hz–18 kHz
Designed for miking strings, pianos, saxophones, hi-hat, and toms.
Try on: strings, snare drum, toms, and dulcimer.

CASCADE FAT HEAD

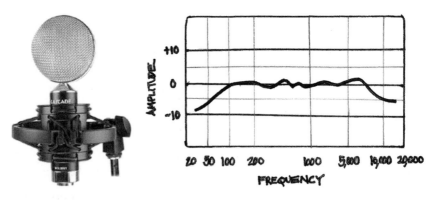

FIGURE 5.80   **FIGURE 5.81**

$
Ribbon
Figure 8
30 Hz–18 kHz
Designed for a Blumlein setup as well as with live application.
Try on: drum overheads, room sounds, guitar amps, vocal, piano, and strings.

COLES 4038

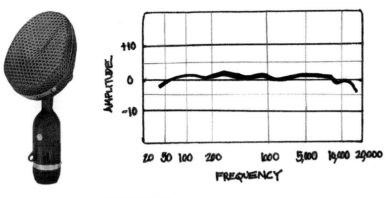

FIGURE 5.82   **FIGURE 5.83**

$$$$
Ribbon
Figure 8
30 Hz–15 kHz
Designed for broadcasting and recording applications. Used by the Beatles and the BBC.
Try on: drum overheads, guitar amps, vocals, and room sounds.

NADY RSM-2

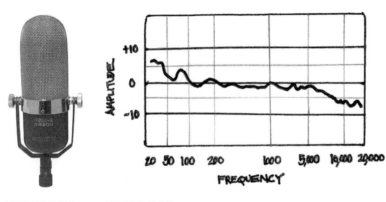

FIGURE 5.84   **FIGURE 5.85**

$
Ribbon
Figure 8
30 Hz–18 kHz
Designed for recording studio vocals, acoustic instruments, ambient sounds, strings, and horns.
Try on: room sounds, drum overhead, vocals, and guitar amp.

ROYER R-122

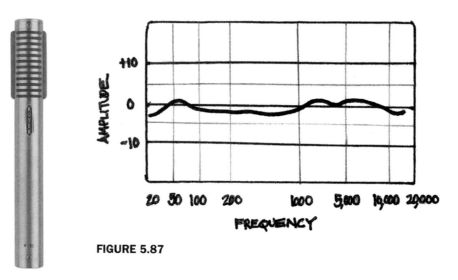

FIGURE 5.87

FIGURE 5.86

$$$$
Phantom powered ribbon
Figure 8
30 Hz–15 kHz
Designed for a wide variety of acoustic and electric instruments, brass, and drum overhead.
Try on: guitar amps, room sounds, drum overheads, acoustic guitar, horns, and strings.

RCA 44

$$$$
Ribbon
Figure 8
50 Hz–15 kHz
Designed for radio and television broadcast announcers around 1932.
Try on: room sounds, acoustic bass, strings, and female vocals.

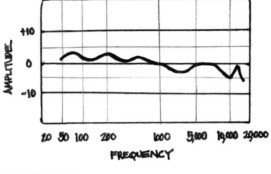

FIGURE 5.89

FIGURE 5.88

SE ELECTRONICS VOODOO VR1

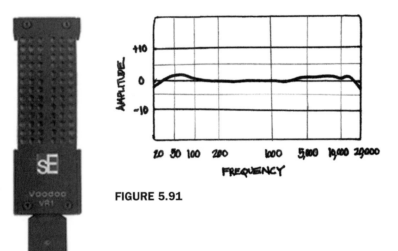

FIGURE 5.91

FIGURE 5.90

$$$
Ribbon
Cardioid
20 Hz–20 kHz
Designed to be one of the first ribbon mics to capture frequencies above 18 kHz.
Try on: electric guitar, percussion, and voice.

CHAPTER 6

Mixing Consoles. So Many Knobs, So Little Time

THE MIXING CONSOLE

A mixing console may commonly be referred to as a mixer, a desk, a board, a soundboard, an audio production console, or by a list of other names. No matter what a console is called they all perform the same basic functions. All consoles have faders that allow for volume control, pan knobs to position a sound from left to right, an auxiliary section to control effects, headphones or monitors, and a whole host of other functions that will be discussed in this chapter. Mastering the console will be necessary to succeed in audio production. Think of the console as command central, controlling audio and its various components. Much like the dashboard of your car that has controls for the A/C, radio, and lights, a console has audio-related controls. Consoles are used in recording, live sound, or in conjunction with digital audio workstations (DAWs). In the all-digital world, there are virtual mixers, but there are also *Control Surfaces* that allow you to work on hardware instead of clicking a mouse when using a DAW.

The Channel Strip

Whether you are working on a physical or a virtual console, many of the controls are the same. If there are 32 channels   **FIGURE 6.1** on the board, understanding the controls on one channel is a huge step because understanding one channel means you will understand all the 32 channels. Once you learn one channel strip, you will know quite a bit about the console or virtual mixer.

4 The Channel Strip

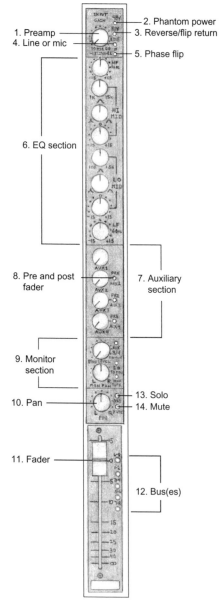

1. Preamp
4. Line or mic
2. Phantom power
3. Reverse/flip return
5. Phase flip
6. EQ section
8. Pre and post fader
7. Auxiliary section
9. Monitor section
10. Pan
13. Solo
14. Mute
11. Fader
12. Bus(es)

Most console channel strips, real or virtual, will include the following functions:

1. **Preamp** – Used to amplify a microphone's signal.
2. **+48 V** – Provides phantom power required by most condenser microphones.
3. **Reverse/flip return** – Usually a function with in-line consoles. This button gives you another option besides the fader to send or return a signal. This return option is often a choice between a *pot* and a fader.
4. **Line/mic** – Determines if the channel will be used with a mic (send) or as a line (return/monitor).
5. **Phase flip** – Flips the polarity 180 degrees on the channel selected. This is a great feature to use when checking phase between stereo mics.
6. **EQ section** – Varies from console to console and with the type of software being used. It can change the tone of the signal that is being routed through the channel.
7. **Auxes/auxiliary** – Used for monitoring and effects (FX). With live sound, they control the mix sent to the stage monitors. In the recording studio, auxes are used to build a headphone mix. When the aux is used for monitoring you will typically select "pre." Both in the studio and with live sound, auxes are used to apply FX to a particular channel. When the auxes are used for FX you will typically select "post." This ensures that the FX isn't heard unless the fader is up.
8. **Pre and Post** – Allows the user to select whether Aux sends are before or after the fader. In "pre," the aux path would be before the fader, meaning the fader would not affect what is being processed. As previously mentioned, "pre" position is typical when the aux is being used for headphones or monitors. When you turn a fader up or down it will not affect the headphone mix. In "post," the auxiliary knob would be after the fader, meaning the fader position would affect the aux send. The "post" position is used when the aux is being used for FX, unless you still want to hear a ghostly FX sound when the fader is all the way down. I have only used this method once in about 25 years, so that should tell you something...

FIGURE 6.2

9. Monitor section – Another option to return the signal from the recorder to monitor the sound. This section is typically used during tracking and can also be referred to as the tape return section or as Mix B.
10. Pan – Allows you to place a sound anywhere between the left and right speakers.
11. Fader – Controls level being sent to a bus or a recorder.
12. Bus(es) – A common path that allows the engineer to combine signals from as many channels as you select. A bus on a console is typically configured for 4-, 8-, or 24-channel mono or stereo L/R. Often the Bus section is shared with the Group or Subgroup section.
13. Solo – Allows you to isolate the channel selected from the others. However, multiple channels can be soloed at the same time.
14. Mute – Turns off sound on the selected channel; opposite of solo.

⚠ TIP

Don't rely on using the solo function too much while mixing. It is more important for music to sound good as a whole, rather as individual sounds. It is not uncommon that an individual sound may not be what you really had in mind but it sounds perfect in the mix. If you rely on solo you may not discover how a sound works with all the other sounds.

Preamp

The preamp controls the gain and is used to boost a microphone signal. The preamp section usually has a +48 V button to apply phantom power to the required mics. The line button is often found in the preamp section. When the line button is pressed down, it receives signal and acts as a tape return. Although many

Preamp section

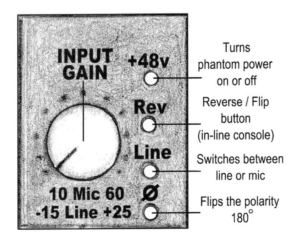

FIGURE 6.3

consoles don't say it, when the line button is up, mic is selected. So really the line button is either line or mic depending on your selection. If the console is an *in-line* type, there will be a reverse or flip button possibly near the preamp to allow the channel to send and receive signal simultaneously. The phase flip option gives the user the ability to flip the polarity 180 degrees on the channel selected. This can be helpful when checking phase between two or more mics.

EQ Section

The EQ section gives control over the tone or color of the instrument being run through the selected channel. Both amplitude and frequency adjustments are made in this section. This allows the user to boost or cut a frequency or range of frequencies. The most common parameters associated with an equalizer are frequency, amplitude, and slope (Q).

Parametric EQ

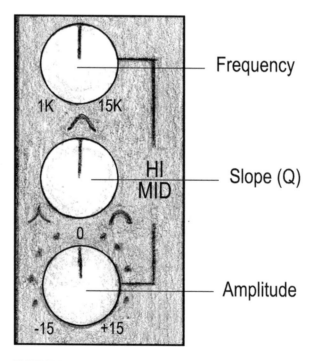

FIGURE 6.4

EQ sections are divided into specific frequency bands. Many consoles will be broken up into three bands with a low-frequency knob, a mid-frequency knob, and a high-frequency knob. Some consoles will have a fixed set of frequencies to choose from, while other consoles give control to select virtually any frequency between 20 Hz and 20 kHz.

Next to each frequency knob will be another knob, or pot, to adjust the amplitude of the frequency you have chosen. There will be a center point on the amplitude knob, where nothing will be boosted or cut. Adjusting the knob to the right of this point will increase the amplitude of the frequency selected. Adjusting the knob to the left of this point will decrease the amplitude.

The slope is the area around the frequency selected. If you have the ability to adjust the slope you are working with a parametric equalizer where slope settings range from extremely narrow, or notching, to extremely wide and gentle. A narrow slope is helpful when notching out noise. Choosing a narrower slope will affect a smaller range of frequencies. A wider slope will affect a larger range of frequencies. Choosing a wider slope may allow you to eliminate unwanted noises, but in doing so you may likely alter or affect your desired sounds as well.

Most EQ sections include a low-cut and/or high-cut filter. The frequency associated with this filter varies from console to console and is often indicated next to the button you are selecting.

More affordable consoles tend to have at least one fixed frequency knob to save on cost. This is usually some type of shelving EQ. With a shelving EQ, the preset or frequency selected is the starting point and not the center point as with many other EQs. These fixed points are either high or low frequencies and when selected all frequencies are boosted or cut the same amount above or below the point, creating a shelf. Some common presets would be a low shelf (LF) at 60 or 120 Hz and a high shelf (HF) at 12 or 16 kHz. With these controls, only amplitude can be adjusted and not frequency.

Shelving EQ

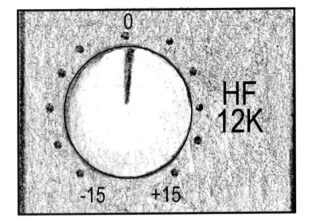

FIGURE 6.5

Auxiliary Section

Auxes are used to control FX sends. They are also used to control headphone and monitor mixes. I have worked on consoles that have anywhere from 2 to 18 auxes per channel. In the studio, auxes are used to create a headphone mix or are used as FX sends. For live sound, auxes are used to create a stage or monitor mix and to control FX. The "pre" button located near the aux send, when selected, puts the aux path before the fader. The pre is selected when the aux is being used as a headphone or monitor send. By this way, the engineer can have their own fader mix and the adjustments won't affect the headphone mix. The pre will not be selected when you want to use the aux as an FX send.

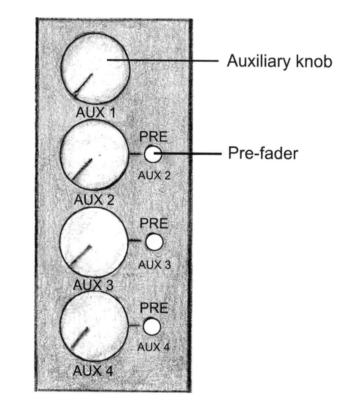

Auxiliary section

FIGURE 6.6

Monitors, Faders, and Busses

Some consoles, especially those known as *in-line*, will allow the user to both send and receive signal on a single channel when recording. If you are using the channels preamp, it is necessary to have both a tape send and a tape return. The fader would typically be your monitor level, but when recording it controls the amount of signal going to your recorder. If the fader is being used to send level to the recorder, you obviously won't be able to monitor the return on that same fader. In-line consoles allow you to send level on a channel and also monitor level on the same channel because of an additional pot or fader. On some consoles, the monitor return is called "Mix B" with "Mix A" represented by the fader. When recording with a *split* console, half the console's faders are used for sending level to the recorder and the other half of the faders are used to monitor the sound. With a 32-input console channels, 1–16 could be used for sending level to tape and channels 17–32 used to monitor the sound (return). Channel 1 would send level to track 1 and channel 17 would be used to monitor the track. Channel 2 would send level to track 2, channel 18 would monitor track 2, and so on.

Fader

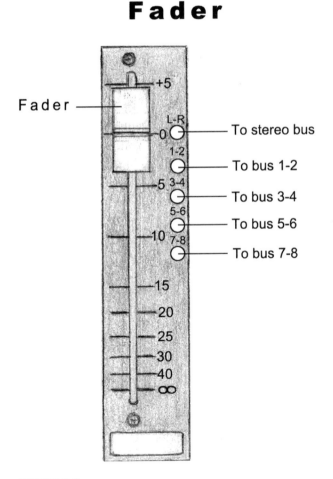

FIGURE 6.7

A fader controls the volume for that channel. It is almost always found at the bottom of the console, and to some, it appears as a "slider." As previously mentioned, if a mic is plugged into a channel, the fader may control the output level sent to the recorder. Faders can act as returns to monitor signal or as sends to control output to a recorder. There will be more discussion about signal flow in Chapter 8.

The bus section can be found near the fader at the bottom of the console or at the top of the console near the preamp. When the signal returns to the channel, it must be assigned to a bus to be heard. The most common selection would be to press L/R or stereo bus on the channel. This would send the signal to the stereo fader, which in turn feeds the speakers and two-track recorder. Besides the L/R or stereo bus, consoles have additional busses. Most consoles have eight busses, but some have up to 24 busses. The bus allows you to assign the channel to a variety of signal paths. If level is returning to a channel and can't be heard it might be because a bus (path) for the signal has not been selected.

Master Section

In addition to the channel strip, recording consoles have a master section. The master section will have the master controls for the faders, auxiliaries, speakers, and other inputs and outputs, like an external two-track or CD player.

The master fader is where all individual sounds are summed to two channels: left and right. The output of this commonly feeds the speakers and any external two-track recorders such as an Alesis Masterlink or ½" tape machine.

The master auxiliary section controls the overall volume of the auxiliary selected. Most consoles have both an Aux Send master section and an Aux Return Master section.

The master solo controls the overall solo volume when you solo a track. I set this to match my overall listening volume. When you solo an instrument, it will sound like your listening volume and you will be less likely to make unnecessary adjustments based on the volume. Often we think louder is better…

Main, sometimes called monitor, volume controls overall listening volume for speakers. This typically controls the volume for the stereo bus, two-track returns, or any other signal played through speakers.

Some additional controls common on an audio production console are as follows: Alt(ernate) Monitor switch, Mono button, and Talkback mic and button. These controls are often grouped together on the board.

The alternate monitor button will allow the user to switch between multiple sets of speakers. Studios typically have at least two sets of speakers that the engineer will use to A/B (compare) sounds. Consider having at least two sets of speakers that specialize in different areas. Make sure mixes sound good on lower quality speakers as well as the best pair. Always remember that if a mix sounds good on a lower quality pair of speakers, it'll more than likely sound good on just about anything.

When the mono button is pressed, the stereo image is summed to mono. I constantly check my mixes in mono to make sure that they are in phase and that they would hold up if played on a mono system.

The talkback mic is useful when the musicians are wearing headphones, using stage monitors, or if there are playback speakers in the studio. Talkback mics are often mounted flush with the surface of the console. Other talkback mics can be attached to the console. The talkback button is pushed to activate the mic. In the studio, the talkback mic signal is usually routed through the aux sends feeding the headphone mix. This makes it easier for the engineer or producer to communicate with the musicians in the studio.

Talkback section

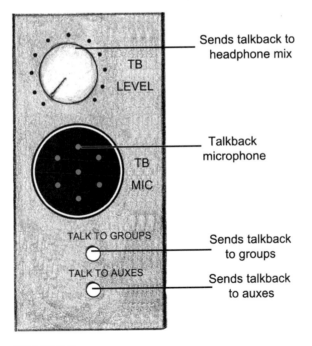

FIGURE 6.8

Meters

Meters on a console indicate the amount of signal input or output from the channel, stereo bus, groups, or auxiliary section. Meters come in many forms, with the most common being a variety of peak and VU meters. Modern audio equipment uses peak meters because they respond instantaneously to an audio signal. This is necessary with digital audio equipment.

VU stands for volume units and this type of meter responds much more slowly than a typical peak meter used with digital audio. Most VU meters will have a needle that measures the average amount of signal. This is similar to how our ears react to sound. This type of meter is common with analog gear.

Peak meters typically have green, yellow, and red LED light indicators. With digital audio, a meter is needed to quickly respond to transients to ensure that no levels exceed zero and square off the sound wave. A peak meter responds quickly and is essential with digital audio because recording levels in the digital world cannot exceed 0 dBFS. Unlike a VU meter that measures the average amount of voltage, a peak meter measures the peaks.

No matter what kind of meter is being used you should always trust your ears first. Believe it or not, meters are not always accurate. If it sounds good, it is

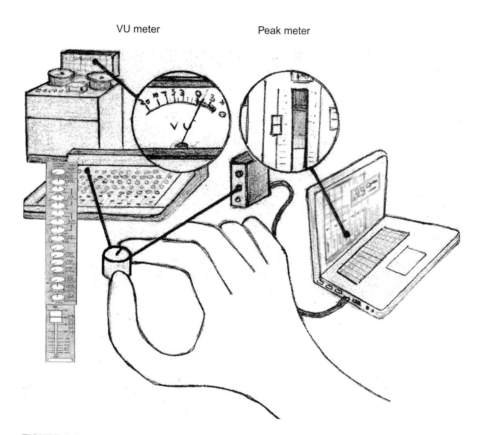

VU meter   Peak meter

FIGURE 6.9

good. Meters are there as a guide and should give you confidence that you are operating your equipment at acceptable levels. Meters are not just found on consoles, they are also found in audio software programs and on most pieces of analog and digital outboard gear.

OTHER CONSOLE FUNCTIONS

Monitoring Sound

Monitors, or speakers, are an important part of any studio. After all, monitors are how you evaluate and listen to a sound. You need to be familiar with both the speakers and the room in which you are working. Although some larger commercial studios still have "Big" speakers, most studios stick to *midfield* and nearfield monitors. In fact, most home and smaller studios have only nearfield monitors. Nearfield monitors are placed on the console bridge or on speaker stands very near the console. They typically have a woofer in the 5"–8" range and a small tweeter. I prefer two pairs of nearfield monitors. One pair that emphasizes the midrange and another pair that scoops out the mids, emphasizing the lows and highs that make it easier on the ears for long listening sessions. You don't need the most expensive pair of monitors to mix successfully. You just need to be extremely familiar with your speaker's frequency response. The same can be said for the room. You don't have to have a perfect acoustic space to mix in but you do have to be familiar with its sound. If you know that your control room cancels out bass frequencies, you won't add more bass because you are familiar with the amount of bass you should hear in a typical mix when mixing in that particular space. Listen to a lot of music in the room you will be mixing in. It will help you get to know how things sound in the room.

It's a good idea to listen to mixes on different speaker systems. In addition to listening on your main speakers, you may want to reference mixes on an old school jam box, headphones, surround sound system w/subwoofer, or your car. The car is a very common place to listen to mixes because most people listen to more music in their cars than anywhere else. Many musicians prefer to hear the mixes in their own cars since they are familiar with the sound quality and EQ settings. This often gives them confidence that the mix meets their expectations. Monitoring on different systems can help you determine if your mixes will translate on multiple speaker set-ups and not just your studio monitors. You may even want to reference your mixes on computer speakers or ear buds since many people now listen to music this way. You would hate for your dance mix not to have kick drum in it when people are checking out a mix online.

For the best, stereo image monitors need to be properly placed. A traditional way to setup speakers is to make a triangle with equal distance between each point and sit at the point of the triangle. This is known as the "sweet spot." Make sure you point the speakers inward at approximately 60-degree angles. Otherwise, the stereo image won't be properly represented. If you set your monitors too close to each other, the stereo field will be blurred. If you set your monitors too far apart, the stereo image will have a hole in the center. In terms of height, try and make sure that your tweeters are at ear height.

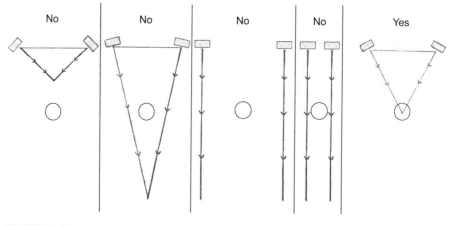

FIGURE 6.10

Patchbay

A patchbay contains all the inputs and outputs of most, if not all, of the equipment in the recording studio. The patchbay allows you to access and redirect signal flow. A patchbay typically contains the console direct outs, the multi-track inputs and outputs, the console line-ins and line-outs, and all the outboard gear (compressors, FX processors, gates, and equalizers). The patchbay usually has tie lines to the studio so you can redirect the mic inputs.

If you are using more than eight tracks, especially if you are using all analog gear, a patchbay will be your best friend. You will quickly realize this if you try operating without a patchbay. Without a patchbay you will have to go behind your console and outboard gear to patch and un-patch any cables. Say you want to patch in a compressor to a particular track, you would have to run a cable from your tape output into your compressor and then run another cable from the output of that compressor into the console line-in. Believe me, this will get confusing and old quickly! With a patchbay you will take a patch cable and make the same thing happen in seconds.

Even if you are working entirely in the box (doing everything with a computer) you may want to consider a patchbay. If your computer is doing all the

Patchbay

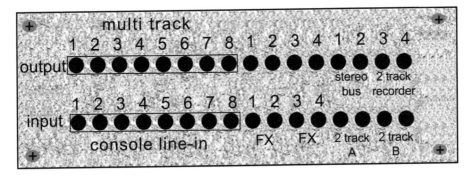

FIGURE 6.11

processing (FX, dynamics controls, and automation), it will take a lot of CPU power to keep up. Your computer will run more efficiently if you use outboard gear to do most of the processing. Plug-ins are great, but the real deal is even better. If you use outboard gear, you will need a patchbay to access this gear or you will be running a bunch of unnecessary cables every time you want to use your prized possessions.

Many studio owners have expressed to me that young engineers have no idea how to make a simple patch. Literally, learning the ins and outs of a patchbay can separate you from other up-coming engineers lacking this skill. Professional studios have patchbays and you will eventually need to learn how to use one.

INTRODUCTION TO MIXING

All Mixed Up

The term "mix" comes up quite a bit when talking about recording. This section will clarify different ways to use this term and introduce you to a few basic mixing techniques.

Mixing (verb) – The art of blending, editing, EQ'ing, and processing the recorded tracks to a stereo, L/R mix. As in, "I have been mixing this track for three days."

Mix (noun) – The result of mixing. A stereo, two-track, L/R mix. As in, "I did a vocal up mix."

Mixed (verb) – Past tense of a mix. Usually means a mix that has been completed. As in, "I mixed ten tracks last week."

Mixer (noun) – The person doing the mixing or the audio-production console. Both can be referred to as "mixers."

The three steps of the mixing process are as follows:

1. Visualization. Picture how you want the mix and the individual instruments to sound. Instead of aimlessly turning the EQ knob trying to find a frequency range that sounds good, visualize how you want the sound to end up. Fatter? Brighter? Edgier? You can then turn the knobs with a purpose.
2. Application. This involves fader adjustments, panning, signal processing, equalization, and applying FX to the mix.
3. Evaluation. Evaluate the changes you just made. Were you able to achieve your vision with the action taken? Evaluate what is working and what isn't and continue this three-step process until everything is tweaked the way you like.

A major difference between the professional engineer and the novice is that the professional can visualize how they want something to sound and quickly apply the necessary tools to achieve that vision. A novice, on the other hand, is more apt to search for sounds with less direction. The more you mix the easier it will become to achieve a particular sound in less time. Practice mixing as much as possible. This is really the best way to improve your mixing skills.

A good mix should

- Allow the listener to focus on what you want the listener to focus on.
- Sound good on *all* speakers, not just a high-end pair.
- Highlight the best elements and mask the weaker elements. Just as a photographer uses lighting to accentuate positive details, a recording engineer can apply this same principle. For example, maybe the backing vocals aren't as awesome as they could have been for one reason or another. Find a way to mask the backing vocals so they don't stick out and/or use an FX to make them sound interesting without too much focus or detail in the overall mix.
- Have a good balance between the instruments, vocals, and other sounds. If the mix consists of a folk band with an acoustic guitar, vocals, fiddle, acoustic bass, and a small drum kit and all you hear is kick drum, probably safe to say this is not a good mix.
- Be appropriate for the song and the genre. This isn't to say you shouldn't break the rules, but you should know that musical genres often have their own style and sound. Make sure you understand the sound the band is seeking. Mixing outside the accepted norm of these genres can make you a hero or it can be a disaster, should you stretch too far or get it wrong. You need to know the rules before you can break them successfully. I encourage bands to give me a few CDs of music they like to listen to weeks beforehand so we can have some reference points later. The band will also respect you for knowing more about their genre of music. It is always a good idea to have the reference

material close to hand when mixing. Although you're not trying to copy it exactly, it will give you a rough idea how it should sound in relation to the genre.

- Have some dynamics. Keeping everything at a loud volume will get old. Not to mention, dynamics are an excellent way to create emotion. When you see a scary movie, the music slowly goes from quiet and soft to loud and shrill, to indicate a rise in the action and keeps you at the edge of your seat. Watching the staircase scene in *Psycho*, with the volume turned off. It is a very different experience. This dynamic experience creates the tension and emotion. Along the same line, dropping out multiple tracks can be a very effective way to make the listener aware and emotionally invested in the song. Don't be afraid to drop out the drums or guitar. Just because it was recorded or always played that way doesn't mean that it has to be in the final mix.
- Make the listener react as the songwriter intended. A dance track should make you want to get up and dance just like an aggressive metal song should make you want to rock out and bang your head.
- Leave the listener humming the melody or vocal line.
- Move you and the listener emotionally. Music is about vibrations and creating a feeling. If you can bring that out as an engineer you have done a great service!

🔺 TIP

Instead of searching for what you want in mix by trying different EQs, reverbs, and other signal processing, visualize what you would like to hear and then apply those tools and tweaks necessary to achieve your vision.

How to Improve Your Mixing Skills?

Analyze mixes of different genres and listen closely how the songs are mixed.

Practice! It takes years and years of terrible mixes to become halfway decent.

Compare other people's mixes to your own. What is different? What is similar? Make notes. "My mixes don't seem to have as much bass." or "My mixes lack dynamics."

Learn how to recognize the different frequency ranges and how they make you feel.

More practice—this cannot be overstated! Becoming a good mixer will take years and years of practice.

Don't let the pressure get to you. There will be situations in the studio that you will have to learn to deal with, like, everyone wanting their instrument turned up the loudest, or someone's friend or relative that has no obvious musical talent suggesting to you how to mix the song. Other pressures may include getting a decent recording or mixing within a short time frame and within the artist's budget and keeping the artist happy. Making the artist happy is tricky and can bring out insecurities as an engineer. Presenting a band mix is similar to them presenting their songs to you. You really want people to like it but you question whether or not it is worthy. The chances are you won't make everyone happy. Maybe the bass player doesn't like his tone or the guitar player wanted his solo even louder in the final mix. Whatever the reason someone doesn't like the final mix, you will need to get used to dealing with the fact that you can't please everyone. Be confident in your decision-making but don't be cocky, defensive, or unreasonable.

Put your heart into it. To most music lovers, music is an emotional experience and often quite personal, so imagine how it feels to have people judging you whether or not you have made a good mix or written a good song. I have been mixing for over 25 years and every time I present a mix for the first time I have butterflies. I try not to show it, but I care what the band thinks. I put my heart and soul into that mix and I hope the band gives me the thumbs up, the metal horns, or whatever moves them to express their emotion.

Don't worry about things beyond your control. There are some things you won't have control over as an engineer, like the song selection. Unless you are also producing the band, the chances are you have zero control as to whether or not the song presented to you is a "hit." A good song is a key component to any good mix. I didn't realize this early on in my career and I would often ask myself why my mixes didn't sound as good as my mentor's mixes. One reason is that the level of talent I was working with was much lower. My mentor was recording experienced songwriters with great sounding instruments. Their arrangements also made it easier to achieve a good mix. Eventually, I figured out that a weak song, with weak players, and weak sounding instruments are almost impossible to make sound good. As you improve your skills, access to record better songwriters and musicians will make your recordings sound even better. As a young engineer you should expect your results to be proportional to the band's talent. However, your ultimate goal should be to make the artist sound even better than their talent level. As an engineer, it is rewarding when someone compliments how great a band sounds and how much he or she enjoys the recordings of a particular artist. Only the engineer knows the truth. It took a lot of hard work to make an average band sound better than average. Regardless of anything else, the reward is in knowing you did a good job and your audio skills serviced your client well.

Mixing is both a technical and an artistic adventure. It will require more than just turning knobs. It isn't something you will learn overnight and it will require both patience and perseverance. Not only will you have to learn the technical side you will have to learn how to deal with people, make hard decisions, and be at your creative peak. In the end, you will need to understand the tools of the trade and also understand how to bring out the emotional components of the mix to be an exceptional mixer.

For more information on mixing check out these two books: *Mixing Audio: Concepts, Practices, and Tools, www.MixingAudio.com* (Izhaki, Focal Press, 2008) and *Zen and the Art of Mixing* (Mixerman, Hal Leonard, 2010).

CHAPTER 7
Signal Processors. Toys You Could Play with for Days!

WHAT ARE SIGNAL PROCESSORS?

Signal processors take an original signal and manipulate it to produce a new sound. The original sound can be blended with the processor or can be run directly into the processor. They can be purchased as plug-ins, pedals, and rack-mounted or stand-alone units. Signal processors can do everything from making a sound more consistent to making a sound appear to be from out of this world. Basically, they are used to treat a sound for a desired effect or control. Compressors, limiters, gates, equalizers, reverbs, and delays are a few signal processors you may run across.

When beginner audio students are turned loose in the control room, they naturally gravitate toward the effects processors. Students tend to be attracted to the time-based effects like reverb and delay, probably because, unlike compressors, effects are easily heard. Scrolling through endless effects units can be mesmerizing. Searching through every possible effect on every effect unit can take countless hours, a sure way to get lost in an audio vortex! Eventually, you find the ones you love and store those to your memory to recall next time you want that special guitar tone or spacey vocal someone requests. You will probably do the same thing with compressors and other signal processors, that is, experimenting with all the effects patches at your disposal. It will just take longer to learn how to control and hear the results of nontime-based processors like compressors.

CONTROLLING DYNAMIC RANGE

5 Compression

Compression helps control the dynamic range of an instrument, voice, or recording. Compression is harder to master and much more difficult to initially hear and understand than the other signal processors discussed in this chapter. The dynamic range is the difference between the softest and loudest sound with an instrument, recording, or mix. A compressor is great for that emo, punk, or hard-core singer who whispers and then screams as loudly as possible. A compressor can help even out a performance or make two instruments react together. This is a useful signal processing tool when you want a vocal to sit more consistently in a mix or make any instrument sound consistent in volume. Because a compressor turns down the loudest sound, it essentially brings up the quietest sound. Be aware of what that quietest sound is... A/C noise? hum? hiss? It could be a noise, rather than an instrument. If the noise is too loud and is destructive to the desired sound, don't compress the signal too hard (aka squash). Squashing the sound will introduce more of the unwanted noise into the desired sound. Compression can also be used to bring up that quiet sound in the mix. Let's say you recorded a band with one mic in the room and the guitar was really quiet and the drums were really loud. Using compression could make the guitar sound louder and the drums quieter, in essence, making the two instruments more even in volume.

Compression is used to:

- Even out the dynamic range of a sound, instrument, or recording.
- Glue two instruments together so they "move" together, such as with a bass and kick drum track.
- Bring up a quiet sound.
- Control a loud sound.
- "Color" the track a particular way. For example, a Tube Compressor could be used to darken and fatten up a bright and thin track.
- Make a creative or different type of sound, like setting a very fast attack and a super slow release to a recorded cymbal sound. This creates a cymbal hit with very little attack that gradually gets louder while emphasizing the "wash," a great effect for some types of music.

▲ TIP

☀ AUDIO CLIP 7.0

Try miking a drum set with one mic in the room. Compress the sound and listen to how the compression can bring up the quiet drums or cymbals in the mix and how it turns down the loudest sounds. Compression can work great on room mics!

Although compression can be applied to any instrument, the most commonly compressed signals are vocals, bass, and kick drum. Many engineers also apply a stereo compressor to the stereo bus to control peaks, color the sound, or glue the entire mix together.

The main parameters on a compressor are the following:

- Threshold or input
- Gain reduction (GR)
- Output or makeup gain
- Attack
- Release
- Ratio
- Link
- Side chain
- Bypass

Threshold controls when the compressor kicks in and reduces the dynamic range.

Gain reduction (GR) indicates how much the dynamic range is being reduced. Usually, there is some kind of meter that will identify the amount of gain reduction.

Output/makeup gain "makes up" the reduction from the compression of the loudest sound with the output, sometimes referred to as the makeup gain. If you reduced the gain −3 dB, setting the output at +3 dB would make up the gain you lost and now the processed sound would have the same apparent volume as the unprocessed sound.

Attack tells the compressor how fast to kick in when the signal reaches the threshold. A fast attack will kill the transient and deaden the sound (drum tom will sound like a box). A fast attack setting will range from approximately 0.01 to 10 ms. A slow attack will let the transient pass, resulting in a much punchier, livelier sound (kick drum beater will be in your face). A slow attack setting will be approximately 50 ms and up.

Release indicates how long the original signal will be held after it is compressed. Although there are no fast rules here, the release is commonly set based on the tempo of the song so that the sustaining of the original signal does not interfere with any new signals. Many engineers use faster release times for faster songs and slower release times for slower songs. In pop music, faster release times are used so the effect of the compressor is not heard. Again, there are many ways to set both attack and release times. You will need to experiment until you become familiar with what different attack and release compression settings produce.

Ratio (2:1, 4:1, 10:1, etc.) determines the amount of output based on the level of input. A 2:1 ratio indicates that for every 2dB over where the threshold is set will result in a volume increase of 1dB. A 4:1 ratio indicates that every 4dB over the threshold will result in a 1dB increase. Ratios of 10:1 or greater are referred to as limiting instead of compression. Limiting provides a kind of "brick wall" effect.

You may also see "soft knee" or "hard knee" when dealing with a compressor.

A hard-knee setting will abruptly reduce the signal's gain as soon as it reaches the threshold setting. This is a much more noticeable setting. A soft knee will gradually compress the signal as it reaches the threshold setting and is a much more subtle setting. Most "auto" settings apply a soft knee.

⚠ TIP

Until you are familiar with compression, I recommend using "auto" settings and not compressing the sound too much (maybe a decibel or two of gain reduction at a 3:1 ratio). Of course, don't be afraid to experiment with different levels of compression and settings, so you will be familiar with the compressor controls when it counts.

Link allows you to join a two-channel compressor or a single-channel compressor to another single-channel compressor for stereo use. You may choose to use this feature when using stereo bus compression or stereo mic setups. In this operation mode, it is typical for one channel to act as the master control for both the channels, subsequently overriding the second channel or second compressor settings.

Side chain allows you to use a signal other than the main input to control the amount of compression. Often an EQ is used, resulting in compression being applied to what is being equalized. The side chain is also used for *de-essing*, multiband compression, and selective gating.

Bypass is used to compare a signal with and without compression. It's a good idea to switch back and forth between the compressed and uncompressed signal. You may find that you don't need any compression after all or that the output will need to be increased (makeup gain) to match the original volume.

⚠ TIP

Some styles of music require and expect more of a compressed sound than others.

Pop, rap, hip-hop, metal, new country, and modern rock typically are more compressed than other styles. Loud and bass heavy music requires more compression to control these more extreme amplitudes and frequencies. A compressed sound is also common with commercial ready music.

Folk, jazz, bluegrass, classical, old country, and most indie rock tend to have a less compressed sound. This helps to retain the natural dynamics and authenticity of these particular styles.

Whether compression is used to control the dynamic range or for special effect, it is a common signal processing tool that will require much interaction and experience to master its results. Unlike EQ, compression and limiting are much harder to initially hear and control.

Gates

A noise gate, or gate, allows the user to turn on and off a sound. If the gate is open, it allows for a sound to be heard. If the gate is closed, no sound will be heard. Gates are typically amplitude based. For instance, a drum mic may be gated so that the mic is only "open" or "on" when you strike the drum. The gate would remain closed until a certain volume or threshold has been met. Like a compressor, the threshold parameter will control when the gate is activated. Gates are most commonly used with drums and in live sound. Some gates have the ability to adjust to frequency as well as amplitude, like the classic Drawmer gates. The opposite function of a gate is known as a "duck." Many gates have this function as well. A duck allows you to bring down a sound when triggered. Ducking is common in broadcast radio to bring down music when a voice-over is present.

 **AUDIO CLIP 7.1**

CREATING SOUND FX

Although some signal processors are used to control the dynamic range, others are used to simulate or recreate an environment or imaginary space. The following effects (FX) fall into the latter category. These signal processors can add delay, reverb, chorus, flanger, harmonization, distortion, and do just about anything to a sound. FXs are time based, using delay times to create the different sounds. Like compressors, limiters, and gates, they are available as plug-ins, pedals, or rack-mounted units. Let's start with reverb.

Reverb

The remainder of sound that exists in a room after the source of the sound has stopped is called reverberation or reverb. The sound is often characterized as an undefined wash of sound created by different delays. Reverb is probably the most common FX used when recording and mixing music. It can be applied to any instrument, vocal, or sound. Traditional reverbs are halls, plates, rooms, or spring reverbs. Modern technology has provided us with a variety of reverbs that simulate all types of real and imaginary spaces. Reverb can help unify sounds together, as well as help separate sounds. For instance, applying a "small hall" reverb FX to all tracks in a recording would glue them together. The effect applied to all instruments may inadvertently make the instruments sound less distinct and muddy the mix. Applying the "small hall" effect on a select few instruments will help separate the affected instruments from the unaffected, or dry, instruments, creating more depth and less clutter in the mix.

AUDIO CLIP 7.2

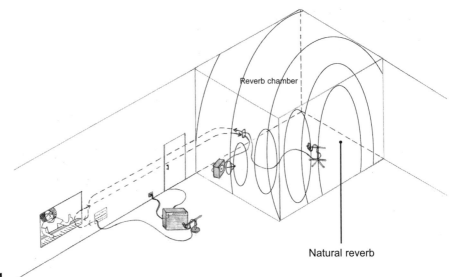

FIGURE 7.1

The main parameters with reverb are the following:

- Choice of FX
- Input
- Mix
- FX balance
- Decay time
- Predelay
- EQ, usually a high-frequency (HF) roll-off
- Bypass

Choice of FX: The first thing to do is choose the FX that you want to hear. Most outboard gear will have a knob or selector button that will allow you to scroll through the FX options. If you are using software, you will also scroll through a list of possible FX and select the one you would like to hear. This could take hours when you are getting to know the various FX at your disposal. As you become more familiar with your FX choices, you will be able to pick and choose much faster. Make notes of the ones you like for future reference. Until you have your list of favorites, prepare to be lost in the land of endless sounds. It will be well worth your adventure!

Input controls the amount of dry signal (unprocessed sound) to be processed. The input signal can usually be adjusted to make sure there is plenty of gain. Be careful not to use too much gain or the signal will be distorted. Generally, there will be an input meter that will help you determine the amount of input coming into the FX.

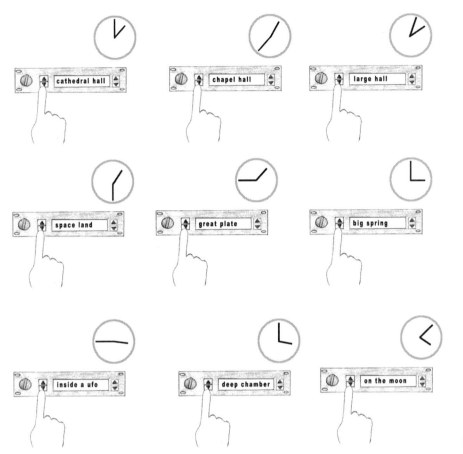

FIGURE 7.2

Mix control usually determines the amount of dry signal (unprocessed sound) versus the wet signal (processed sound) you would like to hear in the mix. It is a balance control between the wet affected signal and the dry unaffected signal.

Decay time allows you to adjust how long it takes the reverb to disappear. Most reverb units have decay adjustments from about 1 ms up to several seconds. Longer decay times make a sound bigger and appear in a larger space. Shorter decay times make things sound more upfront. Make sure the decay time matches the tempo of the music. A long, slow, reverb decay could make the tracks sound muddy or disjointed if the song is fast. Try a longer decay time on a slower tempo song and a shorter decay time on a more upbeat tempo song.

Predelay adjusts the amount of time before the dry signal is affected by the wet (delayed) sound. Think of predelay as the time it takes the dry signal to bounce off a wall or surface before it is affected by the reverb. A longer predelay can help the dry vocal sit more upfront in the mix along with generating a clearer sound that isn't muddied by the reverb.

Bypass allows you to hear the signal with or without the FX in the signal chain. This is basically an on/off function.

▲ TIP

A majority of reverb units allow you to adjust the tone of the reverb or at least offer a high-frequency (HF) roll-off. One way to make a reverb sound less digital and more natural is to roll-off the high frequencies. This is the area where most digital reverb units (especially cheaper ones) tend to sound fake and thin. Try rolling off some high frequencies and note the result.

Delay

A delay is a time-based effect. A delay can take a signal and process it to be played back at a later predetermined time. A delay can repeat a signal once or an infinite number of times. It can be used to separate a vocal from the rest of the mix or for a special effect. It appears natural when used subtly. Apply a large amount of delay and it can make a sound appear spacey and from another world.

The main parameters with delay are the following:

- FX
- Input
- Output
- Time/tap
- Repeat/feedback
- Mix
- EQ

⚜ AUDIO CLIP 7.3

FX: There are a variety of types of delays available. Some delays are straight delays, whereas others offer a variety of delay selections. On a few units, you may only be able to select short, medium, or long delay times. Other delay units will offer delay types such as analog or digital echo, ping pong, delay with modulation, sweep echo, slapback, dynamic delay, tape delay, and many others.

Input indicates the amount of signal entering the processor. Be careful not to overload.

Output indicates the amount of signal exiting the processor. This is typically the blend of the dry and processed signal.

Time/tap allows you to adjust the amount of time between the dry, unprocessed signal and the wet, processed signal. Some newer delay units, especially pedals and rack-mounted FX, provide a "tap," a button, to determine the delay time. This is extremely handy when setting a delay time in a live situation or when seeking a particular delay rhythm.

Repeat/feedback controls how many repeats of a delayed sound you desire. It can be adjusted to repeat anywhere from one to an infinite number of times. Increasing the repeat/feedback to the maximum amount creates a unique sounding feedback loop.

Mix controls the blend of the dry and wet signals.

EQ: Different processors give you different EQ options. The most common is a high-frequency cutoff.

⚠ TIP

How to make an echo or delay sound in time:

Divide 60,000 (number of milliseconds in a minute) by the BPM (beats per minute) to create a synchronized quarter-note delay.

For example:

60,000 divided by 100 BPM = 600 ms.

Therefore, a 600-ms delay would result in a quarter-note delay.

For a half-note delay, double the result = 1200 ms

For an eighth note delay, divide the result by 2 = 300 ms

For a sixteenth note delay, divide the result by 4 = 150 ms, a kind of slapback effect useful for dance music.

Another delay type is an echo, which should not to be confused with a delay. Echo is a classic FX that creates a repeating signal that becomes progressively quieter over time. A slapback echo is used to recreate the sound of a nearby echo source such as a gym wall or canyon. It is created with an unmodulated delay time between about 50 and 150 ms. A popular 1980s effect often used with Dub Reggae, but like any FX, it can be used on virtually anything.

More Time-Based FX

The parameters of the following FX will be similar to those of a delay or reverb. Any exceptions will be pointed out.

Most of the following effects differ because of delay times or because the signal is either modulated or unmodulated. If a signal is modulated, its signal is constantly swept at a predetermined or adjusted speed. A modulated sound sweeps through a specific frequency range. If a signal is unmodulated, there is no sweeping effect or whooshing sound.

Chorus generates a detuned double of the original signal that modulates in pitch slightly over time. Essentially, it is an FX where the original signal is mixed with a modulated delay of itself creating a "chorus" effect. The delay times are generally between 10 and 60 ms. This can happen naturally with groups of strings and vocals. Chorus is often applied to vocals, guitar, bass, keyboards, or strings to make them sound more in tune, thicker, and dreamier than a single sound, while still sounding more or less like a single sound. A chorus FX can come in a pedal form, rack mount, plug-in, or be made naturally by recording multiple takes of a single instrument and mixing them together.

Chorus FX may provide additional control over the following:

Depth: The amount and intensity of chorus and/or modulation.

Sweep: The speed and/or range at which the chorus modulates.

Speed: Adjusts the rate of the LFO and pitch modulation.

 **AUDIO CLIP 7.4**

TIP

The next time a vocal or group of vocals sound out of tune, try a little chorus and note the result. I find this as a great FX, especially for out-of-tune strings or vocals. Even if the source isn't out of tune or you don't have a chorus FX, you can have a singer, guitarist, horn player, or string player double or triple their part. This often creates a natural chorus FX for a lusher and thicker sound.

Doubler is used to approximate the sound of two instruments or vocals performing in unison. Most commonly heard on vocals, this doubling effect is created with unmodulated delay times between 40 and 90 ms.

AUDIO CLIP 7.5

Flanger creates a "whooshing" or "jet-like" sound. It is created with a delay setting of 1–25 ms. The delay is then modulated by a *Low-frequency oscillator* (LFO). The LFO setting determines the rate and depth of the effect. There are several claims on when it was first used, word is that John Lennon apparently coined the term "flanging" when referring to George Martin about this effect, actually calling it "Ken's Flanger" naming it after one of Abbey Road's Engineers.

AUDIO CLIP 7.6

Phaser causes various frequencies in the original signal to be delayed by different amounts causing peaks and valleys in the output signal. These peaks and valleys are not necessarily harmonically related. This effect sounds very similar to a flanger. Flanging, on the other hand, uses a delay applied equally to the entire signal. This is similar in principle to phasing except that the delay (and hence phase shift) is uniform across the entire sound. The result is a comb filter with peaks and valleys that are in a linear harmonic series.

AUDIO CLIP 7.7

Harmonizer creates a harmony with the original signal.

AUDIO CLIP 7.8

Overdrive or distortion can make a "clean" sound "dirty." Overdrive or distortion can make a sound fuzzy, crunchy, edgy, and even retro sounding. Most common with guitar effects pedals, however, distortion or overdrive can be applied to any sound.

AUDIO CLIP 7.9

▲ TIP

I like to blend distortion into a recorded snare drum track. Try applying this effect next time you need a snare drum to cut through a big rock mix or you just want to give it a unique sound.

PLUG-INS VERSUS THE REAL DEAL

Plug-ins are a set of software components that add specific abilities to a larger software application such as Pro Tools or Nuendo. It is likely that you have already used plug-ins with other computing activities. Adobe Acrobat Reader, Quicktime, Real Player, and anti-virus software are just a few types of plug-ins that you may already be using. Many people are curious if plug-ins are comparable to the real deal. The real deal being the hardware or original rack-mounted version of the compressor, limiter, or FX unit. Although plug-ins mimic the controls and look similar to these classic pieces of gear, they do not sound the same. This is not to say plug-ins aren't valuable and effective, but they rarely sound like the original versions, with real transformers, tubes, and discreet electronics that warm up and color the sound. These classic pieces of gear can still be found used, as reissues and new.

There are a few advantages to owning audio plug-ins, instead of hardware. You basically own an unlimited amount of copies of a particular effect. They are compact. With plug-ins, you don't need a bunch of extra space in the control room for racks of gear. They are more affordable. While you could maybe only afford one real LA-2A limiter (a classic piece of outboard gear that isn't cheap), you can own the LA-2A plug-in, along with a load of various other plug-ins. Owning an LA-2A plug-in is like owning an unlimited number of LA-2As. Well, kind of… like I mentioned, they don't necessarily sound as good, or even close to the original, but they are getting closer every day. Some prefer plug-ins because they are cleaner and generally provide less noise. The settings are easy to recall and automate. Another cool advantage of using plug-ins is when you get a chance to operate the real deal, you will already be familiar with its knobs and controls.

Plug-ins are not just designed to mimic classic pieces of gear. Plug-ins can do all kinds of things unique to the virtual world. Plug-ins can be used to help with restoration of old records by removing pops and other anomalies to specializing in mixing and mastering. My favorite plug-ins change the sound in interesting and unique ways and don't necessarily copy classic gear. The future with audio plug-ins is limitless and exciting.

It doesn't matter if the signal processors you use are real or virtual. These are just tools that can make your job easier or more difficult if you don't know what you are doing. At some point, you will need to learn how to operate and manipulate sound with these tools. Like much of audio production, it can take you years to figure out what the heck you are doing. Be patient, learn the tools of the trade, and forge ahead.

CHAPTER 8

Signal Flow. The Keys to Directing Audio Traffic

SIGNAL FLOW

What Is Signal Flow and Why Is It So Important to Understand?

A key component of audio engineering is to fully understand the analog model of signal flow, that is, how sound travels from the beginning to the end of its audio path. Much of audio engineering is troubleshooting different issues that occur during music production. Understanding signal flow can help you quickly pinpoint and solve a problem in order to keep a session running smoothly. When recording, the signal flow path starts where you plug in the microphone and ends at the speakers, or two-track recorder. Once this concept is understood, you will be able to hook up most gear, understand a patchbay, and troubleshoot yourself out of just about any mess! Knowing the analog model of signal flow will translate to almost any audio situation. Don't get frustrated if you don't fully understand the concept at first. It usually takes some experience and practical application to totally grasp signal flow.

To understand signal flow you will need to review and become familiar with the following terms:

Pre-amp
Mic
Line

Input ready
Tape send
Tape return
Stereo bus
I/O
A/D
Speakers/monitors

There are many ways to record music. Music can be recorded with a separate mixer and multi-track recorder, or on a laptop with an audio interface. You can even purchase an all-in-one mixer and recorder like an old school cassette, or digital four-track. Some engineers use a MIDI controller and software to create songs. It is even possible to record the band live from the soundboard. The following chapter will help you understand how to hook up your gear and figure out exactly what is going on along an audio path. The analog model is also valuable in understanding many aspects of digital recording and can be applied when running or recording live sound.

Recording involves three specific stages. The first stage is capturing the sounds with a mic. The next stage is storing the sounds in a recorder/line. The third stage is hearing the stored sounds on speakers or headphones. Always keep these specific stages in mind when you are recording.

Setting Levels and Gain Structure

The preamp is the first link in a recording chain when a microphone is used. Preamps sound best when run (turned up) between 10% and 90% of their potential. The sound quality is not all that good at maximum or minimum settings. If the preamp volume is set all the way down when recording a loud instrument, it will be necessary to pad the mic or preamp. This padding will allow the preamp to be turned up. Conversely, if the preamp is turned up all the way when recording a quiet instrument, a more sensitive mic will be needed in order to turn down the level.

Be aware of the input level sent to a digital recorder. If the level is too hot, the signal will clip in a displeasing way. Keep the levels below 0 dBFS. It is not imperative that the level be near zero if recording at 24-bit. Strive for setting recording levels somewhere between −20 and −6 dBFS. This will allow headroom for volume spikes, insuring that you won't damage the integrity of the audio signal. If using an analog recorder, store signals much hotter to increase the signal-to-noise ratio. Levels in the analog world can exceed zero and are typically set from −3 to +9.

Some audio devices or console settings are said to have "unity gain." A device that has unity gain neither amplifies nor attenuates a signal. Most signal processors have unity gain; therefore, when a compressor or other device is inserted into an audio system, the overall gain of the system is not changed.

Analog-to-Digital Converters (A/D) and Audio Interfaces

A recording studio's collection of gear often includes both analog and digital components (learn more about analog and digital audio in Chapter 11). An analog-to-digital converter, abbreviated as A/D or D/A converter, allows these different devices to interact. These are usually rack-mounted units, but can also come in the form of a sound card or audio interface.

An A/D converter is an electronic device that converts an analog signal into its digital equivalent. The D/A converters do the opposite by converting digital information back into analog form. However, all converters perform both A/D and D/A functions. A/D converters come in varying price ranges and configurations. Many professionals would agree that the quality of the converter makes a huge difference in the quality of the sound. Speed and accuracy are two of the most important characteristics to consider when you are looking to purchase a good converter. Also, the better the converter, the less likely you will encounter problems such as q*uantization* and *jitter*. Some A/D converters perform only A/D functions and do not include other options.

Audio Interfaces include A/D converters, pre-amps, and other record/mix controls. Audio interfaces are often paired with dedicated software, but many interfaces are not proprietary and can work with any software type. Pictured below are the two-channel Focusrite Interface and the TC Electronic Konnekt 6 Audio Interface

These particular units have mic preamps, phantom power, headphone outputs, monitor controls, USB 2.0, Optical ins/outs, MIDI ins/outs, S/PDIF ins and outs, and additional I/Os. Both of the aforementioned units are affordable options. Other brand name audio interfaces include: Apogee, Lynx, M-Audio, Behringer, Mackie, Digidesign, Presonus, and MOTU. The price ranges are anywhere from $50 a channel up to thousands per channel.

FIGURE 8.1

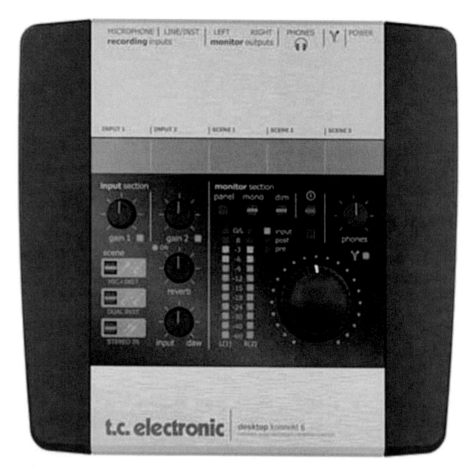

FIGURE 8.2

Analog Model of Signal Flow

6 | Signal Flow

When using a separate console and multi-track recorder signal flow can be represented by the following stages: Capturing sound, storing sound, and monitoring sound.

1. **CAPTURE.** First, plug in the microphone.
2. Turn up mic (preamp) and make sure that track(s) are armed and in input mode on the recorder.
3. Send signal to an assigned track on the recorder. It doesn't matter if the recorder is real (multi-track tape machine) or virtual (digital representation such as Pro Tools). The fader often controls the amount of signal being sent to the recorder when a physical console is used. If a fader is not involved, the pre-amp output will dictate the amount of signal being sent to the recorder.

4. STORE. Now that the signal has arrived at the recorder, the signal can be stored.
5. The signal will return from the recorder to the console, often to a fader or monitor section.
6. At this point, the signal usually has to be assigned to a bus to be heard. Most often the signal is sent to the L/R or stereo bus.
7. MONITOR. Once the signal arrives at the stereo bus fader sound should be heard if the speakers are on and turned up.
8. The signal from the stereo fader may also be sent to a two-track recorder during mix down.

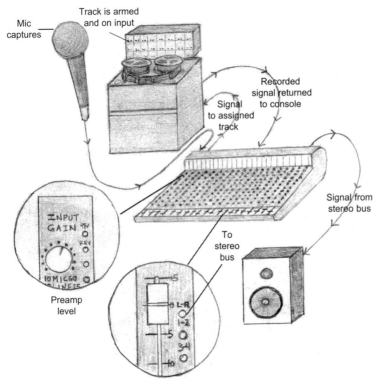

FIGURE 8.3

This path will never change when recording. The only thing that changes is what is being used for the tape machine/recorder. When mixing, steps 1–4 are eliminated and the audio path would start at step 5. When running live sound, you need to only feed the signal to the speakers to monitor.

Additionally, each one of these steps may be represented on a patchbay. If they are, here are a few ways they may be labeled:

- (CAPTURE) Mic tie line, mic input
- Mic output

- Preamp in/out or direct out or tape send
- Preamp out or direct out or tape send
- (STORE) Multi-track in, tape in, Pro Tools in (or any other software-in)
- Multi-track out, tape out, Pro Tools out (or any other software-out)
- Fader in, console or line in, tape return
- Bus out
- Master fader in, stereo bus in
- (MONITOR) Speaker out

Three levels of power correspond to the three stages of signal flow:

1. Mic level (CAPTURE)
2. Line level (STORE)
3. Speaker level (MONITOR)

Between each stage, an amplifier is required to boost the signal to the next power level.

Just like a microphone, a pre-amp can color the sound. For instance, tube pre-amps are popular with digital recording because they can make a thin, clean sound darker and thicker.

When you plug in a mic, you will likely be presented with the choice of "mic" level or "line" level. For a microphone, choose "mic," and for an instrument or tape return, choose "line."

Eventually, the signal works its way down the channel strip and is sent to be stored as a recorded track. This recorded track is stored at line level.

The stored signal is then returned to the console, or fader, to be monitored.

Monitoring is the final stage in the signal's path, and can be achieved by listening to speakers. Speakers are either self-powered or need an amplifier to boost the audio signal. Once the signal is boosted, it is at speaker level. This is the end of the signal path where we can relax and listen back to all of our hard work.

When recording, always picture the three main stages: capture, store, and monitor. Mic, line, and speaker are the hardware equivalents of these stages, respectively. This helps determine where the signal is and where it can be interrupted, if necessary, for manipulation and processing. Knowing these stages and corresponding levels will also make troubleshooting much easier.

APPLYING SIGNAL FLOW
Recording with a Laptop and Audio Interface

The analog model of signal flow directly relates to using a stand-alone recorder with a separate console or mixer. Understanding the analog model of signal flow is useful when working strictly *in the box* or with other recording setup configurations.

When recording instruments or vocals, signal flow always starts with plugging in the mic and adjusting the mic preamp.

Here is an overview of signal flow when recording with a laptop/computer audio interface:

(CAPTURE) Plug the mic into a preamp or audio interface such as an *MBox*. Make sure that a cable is connected between the audio interface and the laptop. Most interfaces use a USB 2.0 connection (see connectors Chapter 15).

Turn up the mic at the preamp.

(STORE) Send the signal to a recorder/computer. Make sure that the correct input and output is selected (often this is IN 1 and 2 and OUT 1 and 2/ stereo bus selected in the software).

Return signal from the recorder/computer. This means selecting the correct bus or output.

(MONITOR) Turn up the monitor or main output knob on the interface.

If speakers or headphones are plugged in and turned up, the signal will be heard.

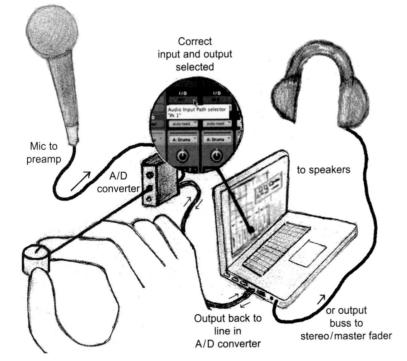

FIGURE 8.4

The two main differences between using a laptop/audio interface to record and using the traditional console/recorder setup are

1. The audio interface's preamp captures the sound instead of the console's preamp.
2. The computer stores the sound instead of a separate, stand alone analog or digital recorder.

Here are step-by-step details that can be followed when recording with most software types:

1. Turn on the computer and audio interface.
2. Open the software application.
3. Set the session parameters, bit depth, sample rate, and audio file type.
4. Determine the file name and destination where you will save the file. Very important! Pick an easy-to-remember location and be consistent with file naming so that you can retrieve files quickly.
5. Create a new track and name it.
6. Put the track into record pause (input ready) by pushing the red button.
7. Make sure that the correct input is selected.
8. Plug your mic into the interface and adjust the preamp. If no mic is required, import the audio track to a desired location.
9. Make sure that the level isn't set too high or too low. Most interfaces will have a meter to check the mic level.
10. Record the track.
11. Unarm the track (take it out of input/record ready).
12. If you can't hear the recorded sounds, select the correct output. Most default settings direct the output to the stereo bus or OUT 1 and 2.
13. Turn up the monitor/volume and listen back. Monitor/volume adjustment is typically found on the audio interface.

⚠ TIP

Make sure that you properly label and save tracks so they are easy to locate and identify. Don't leave audio tracks labeled Audio 1.0, Audio 1.1, etc. Many engineers label tracks by the mic or specific instrument that was recorded. Also, don't forget to back up your audio files to an external drive or other additional storage device.

How to Record Using an Analog or Digital Four- or Eight-Track All-In-One Recorder

FIGURE 8.5

Most analog and digital four-tracks pack a complete studio, minus the mic and speakers, into a powerful little recording unit. These all-in-one units provide multiple mic inputs with preamps. They also provide controls that allow you to adjust EQ, FX, and the volume of each track.

With an analog four- or eight-track you will likely record to a cassette. It certainly is not cutting edge, but it is an affordable way to learn about recording. Fostex and Tascam make classic units that provide a unique quality of sound. These units are completely portable and, if properly maintained, can last a long time. Use high-bias cassette tape and clean the tape heads regularly with 99% alcohol. Cassettes are still available for purchase online.

With a digital four- or eight-track, you are likely to record to a hard disc or flash drive. Some of the routing may be done digitally, but the concept is the same. Digital units are also likely to have some type of two-track built in to bounce down your mixes to stereo, like a CD recorder.

Here are step-by-step details that can be followed when recording with most all-in-one units:

1. (CAPTURE) Plug in the mic. Make sure that mic, not line, is selected.
2. Assign the track to a bus. You may have to use the pan knob. If you were to select the 1/2 bus, panning all the way left would select track 1, while panning all the way right would select track 2. With busses, if you pan all the way left, it will assign the signal to the odd numbered tracks. If you pan all the way right, it will assign the signal to the even-numbered tracks.
3. Turn up the mic at the preamp (likely on the top of the channel).
4. If the correct bus is assigned, the fader is up, and the track is armed, you should see the level at the stereo fader or meter. Make sure that the stereo fader and monitor volume are up at this point in order to hear the signal.
5. (STORE) Record the signal when you are ready.
6. (MONITOR) To listen back to the recorded sound, you will likely flip the channel back to "line" and unarm the track.

With most analog recorders, you will need a two-track recorder for mix down, such as an Alesis Masterlink, or stand alone CD recorder. This will allow you to combine the four- or eight-tracks down to two-tracks (stereo left and right), and store them.

Recording a Live Performance

Recording a live performance can be tricky. In a live recording situation, there is typically very little control over acoustics and a host of other issues may arise. You also may have to piggyback off another system or engineer's EQ settings, which may or may not be sufficient.

One of the easiest ways to get a live recording is to take a pair of stereo outputs from the main mixing console, known by engineers as "getting a split off the board."

Stereo outputs can be recorded to a variety of two-track recorders or straight to your laptop. One potential issue is that your mix will end up being whatever

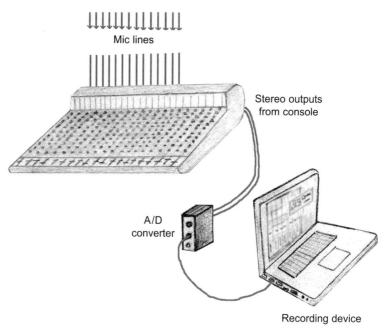

Mic lines

Stereo outputs
from console

A/D
converter

Recording device

FIGURE 8.6

the house sound person has mic'd. A smaller club or venue may not mic the instruments because they are loud enough for the room. As a result, your mix will most likely end up vocal heavy (vocals will be mic'd), and be lacking louder instruments (electric guitar, bass) that were not mic'd by the sound person. Another potential issue is that you are at the mercy of the venue's sound person to set the levels properly.

Another way to record a performance is to place a pair of stereo mics in a good spot and record to your laptop or stereo recorder. The problem with this method is in finding an optimal spot in the room for mic placement. If the mics are placed too far away, you will pick up more of the unclear ambient wash and less of the direct, clear audio. If the mics are placed too close, you may not pick up all the instruments equally. Show up early and scout out the venue or club prior to recording.

🔺 TIP

Try using the XY stereo miking technique when recording a live performance. This technique was discussed in Chapter 5 and demonstrated in Video 3.

Many engineers bring their own mobile recording rigs to record live performances. A few engineers have mobile trucks or vans set up as complete studios. There are even a few folks out there recording live music out of the back of their cars with minimal recording equipment.

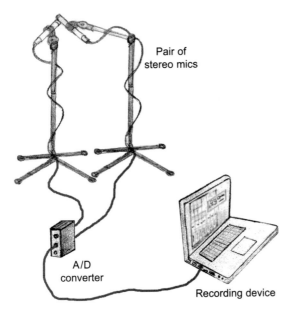

Pair of
stereo mics

A/D
converter

Recording device

FIGURE 8.7

Frequently, people record live bands straight to laptop computers. Keep in mind that this is a live situation. A computer may crash or lockup, which can be a very stressful situation without a backup recorder.

Here are step-by-step details that can be followed when recording a live performance to your laptop or other recorder:

1. (CAPTURE) Place stereo mics in a good spot to capture the overall sound.
2. Plug mics into a preamp or audio interface.
3. Assign mics to inputs and ready tracks by placing them into input or record pause mode on your recorder or laptop.
4. Turn up the mics and monitor the signal.
5. (STORE) Record the signal when ready.
6. (MONITOR) Unarm tracks and listen back.
7. Headphones, not speakers, are used to monitor in live situations.

Venue management, the house sound person, or various setup limitations will determine the location where you will monitor and record sound. Common areas to setup are next to front of house (FOH) sound or to the side of the stage.

When recording a live performance, make sure that you stay out of the way of the other professionals working on the production. Try to be as transparent as possible. Do not take over the house engineer's space or adjust the house engineer's controls. Most of the engineers are willing to work with you if you are courteous and don't overstep your boundaries in their venue.

If you do not completely grasp signal flow at this stage, do not be discouraged! You may initially have to mimic what you have been taught to get a session going. Eventually, you will get the picture. Until then, memorize the steps and any patching needed to record, overdub, and mix. Listen to and observe others that have more audio experience. When you do get to the point where you understand exactly how audio is traveling between any two points, you will have learned an important concept that will help propel you to the next level.

CHAPTER 9
Studio Session Procedures: How a Recording Session Happens and in What Order...

SEVEN STAGES OF RECORDING

Although every recording session tends to be different, most recording sessions will include these seven stages:

1. Preproduction
2. Setup
3. Basic or rhythm tracks
4. Overdubs
5. Rough mixes
6. Mixing
7. Hopefully, some mastering!

It is important to be familiar with the different stages of recording so you can mentally prepare for what comes next. In Chapter 4, I talked about time management and scheduling. You can be better at both of these skills if you know these seven stages inside and out.

So, where do you start?

As an engineer, you will probably start with the setup. However, it is advisable that musicians start with preproduction.

Preproduction

As an engineer, you generally aren't involved during preproduction. I realized early on in my career that you should at least make the recording artist(s) aware of this stage. It will make the session run more smoothly and likely make the overall recordings better. If a recording session doesn't include a producer and the band is inexperienced, the engineer may get stuck producing the session. If this happens, make sure the band or artist understands the importance of pre-production and what it involves.

Preproduction for the band or artist includes picking the songs for the session, working on the songs, evaluating and working out tempos, instrument repair and maintenance (making sure their instruments work!), rehearsing, recording rehearsals or live shows, discussion of overdubs and mixing, and anything you do before the actual recording takes place. If the artist is better prepared, it can save the artist some cash too.

First, make sure the artist(s) have actually rehearsed before they show up to record. You would be amazed how many bands want me to record them immediately, but didn't even know their own songs once they are in the studio. Encourage the band to record a rehearsal or live performance.

Often, artists record an album and months later think how much better they are now. Frequently, they wish they would have been this good when they originally recorded the songs. One reason is because this is the first time they heard the songs recorded and now they know what they actually sound like. If the artist(s) would have recorded these songs first and evaluated them beforehand for things like tempo and song structure, they may have avoided the feelings of regret.

▲ TIP

When evaluating a live performance or rehearsal, consider some of the following:

- Does the song drag or rush in tempo?
- Should it be faster?
- Should it be slower? Is the singer squeezing in all the words?
- Does the drummer need a click track or does he or she just need to practice more?
- Should there be a different arrangement? Just because the band has always played the song in a particular way, doesn't mean it is the best way. Don't be afraid to explore the song with a different arrangement or…
- Should it be recorded in a different key? Does the singer sound like he or she is struggling to sing the song? It is not uncommon for songwriters to change the key to fit the singer's best vocal range.
- Does it sound like everyone is playing together? It isn't unusual for musicians to get lost in their own part and not pay attention to what the other musicians are playing. A recording of a performance or live show will allow the musicians to listen to the overall sound.
- Also, could an instrument drop out and help create a nice dynamic, like the classic breakdown in hip-hop, rap, R&B, and various styles of dance music. If everything is on and up the whole time in a mix, the song may become tedious and will likely lack as much emotion as a song that builds and is dynamic.

These are just a few of the questions you should ask.

Ultimately, preproduction will make your recordings better and along the way save the band time and money.

Preproduction details the band or artist should consider:

1. Set tempos: The songwriter should establish tempo. If you leave this up to the drummer, it may not be the tempo needed for the vocalist. If the song is too fast, the lyrics may seem forced or rushed. If the tempo is too slow, the lyrics may drag and the singer may have to hold out notes longer than they would like. A click track, or metronome, may be needed to stay in time; however, make sure the drummer practices with one first. The click track is fed to the drummer's headphones. BEWARE! Playing to a click isn't easy if the musicians haven't played to one before. A click can be a metronome, keyboard sequence, drum machine, or anything that keeps time. Tempo is measured by beats per minute (BPM).

2. Pick songs: Do this before you show up to the studio to keep the band drama to a minimum and to help focus the recording session. Of course, have some backup songs or secondary songs to record if you have time or in case the songs that were picked just aren't working.

3. Song structure: Work out arrangements and settle any disagreements around song structure upfront. You want to avoid any potential arguments, preproduction can get these awkward moments out of the way. Listen for parts that you could eliminate where there is nothing or too much going on. Be objective! Is the solo section too long? Do you need a bridge? Can the thirty-two bar intro be cut in half? How about the ending? How can you build the song to have more impact? Should some instruments stop playing or dropout during the verses to create dynamics?

4. Rehearsal: The easiest thing an artist can do before they record is to actually rehearse before the session. This will make their budget more predictable and will surely make the recording better! Why pay for studio time when you can rehearse for free at home or for a much smaller fee at a rehearsal room? Again, rehearsing will save time, which equals money in the band's pocket!

5. Record rehearsals or live shows: This will help the band identify spots that can be improved or changed. Use these recordings to help with tempo, song selection, song structure, and anything else you hear that needs work. These live recordings don't need to be high fidelity or professionally produced. Use what you have at your disposal to record a performance for evaluation. The purpose is to hear the song and not to be concerned with the audio quality.

You may want to suggest these things to more inexperienced bands. Most experienced musicians are already aware of preproduction. They understand that being prepared will save time, money, and will likely make the recording better.

Setup

This next stage is where the engineer usually begins. Setup involves tuning drums, addressing isolation issues, picking mics, preparing headphone mixes, starting track sheets or templates, labeling the console, tracks or inputs,

TITLE

NO.	1	2	3	4	5	6	7	8
TRACK	KICK	SNARE	DRUM TOP	DRUM SIDE	DRUM ROOM	BASS DI	BASS AMP	GUITAR CLOSE
COMMENT	D-112	57	M-39	M-39	R-122	DI	EV-RE20	57
EQ / OB								

TITLE

NO.	9	10	11	12	13	14		
TRACK	GUITAR ROOM	KEYS L	KEYS R	DJ L	DJ R	VOCALS		
COMMENT	M-160	DI	DI	DI	DI	TLM-103		
EQ / OB								

TRACK SHEET PAGE ☐ OF ☐

PRODUCTION / PROJECT			ENGINEER	
CLIENT			ASST. ENGINEER	
PRODUCER			JOB #	
REEL #	OF	SPEED	TAPE	IPS
TRACKS		NOISE RED.	☐ ORIGINAL ☐ SLAVE	CAL TONES

FIGURE 9.1

getting levels, and other miscellaneous duties needed before actually pressing record.

Mic selection: I typically show up about an hour before a session and get everything prepared. It is much easier to do this when the band isn't around and you can take your time. Many studios will allow thirty minutes to an hour for setup and tear down.

Setup details are as follows:

1. Isolation issues: You probably will know the instrumentation you will be working with in the session beforehand. I often draw the setup to get a visual and determine where to place the amps, the drums, the vocalist, etc. See Figure 9.2. The purpose is to provide isolation for individual instruments and also to retain a line of sight between musicians, so they can communicate. If you are working with an electronic musician, you will probably set up in the control room and take direct outputs, so you will not have isolation issues. See Figure 9.3.

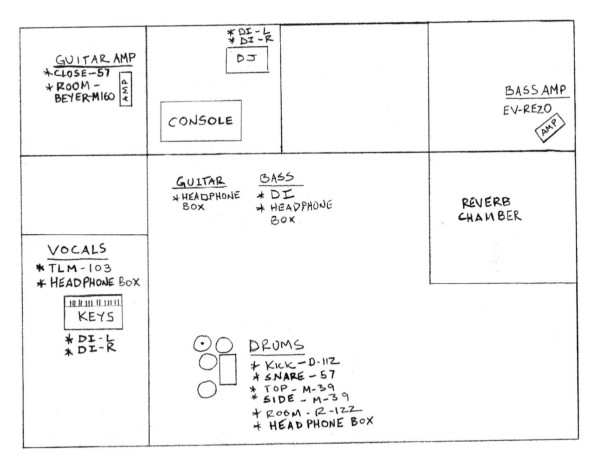

FIGURE 9.2

FIGURE 9.3

2. I will pick the mics I plan on using for the session before the band sets up. Just because you pick a particular mic doesn't mean you can't change it. The point is to make sure everything is working before the band shows up. If the mic you choose doesn't sound good, at least you already have everything patched in and ready to go. Most studios will have their microphone selection on their website or can send you a list upon request. If no mics are being used, you would likely be taking the output of the instrument and plugging it into a direct box or ¼" input on a preamp. You can then test with a drum machine, keyboard, or any other instrument.

3. Get levels: This is much easier to do when you don't have musicians looking over your shoulder. Since you have already plugged in and set up mics, you can apply phantom power to the necessary mics, turn up the preamps, arm the tracks, and make sure everything is working. You can snap your fingers in front of the mic or pump some music into the studio and look at your recording device to make sure you are getting a signal. Again, there is a lot less pressure to do this before the band arrives.

4. Labeling: Label all the gear that you will be using during the session. If you are using outboard gear, label what is going to be patched into it. Also, label the console and fill out any necessary tracks sheets. If you are working on a computer, pull up a template or at least label all the tracks. Don't leave the tracks reading Audio 1, Audio 2, etc. Later this will make your life very difficult! What if you sent your tracks to another engineer or opened them up at another studio and they weren't labeled? It would take a good amount of time to decipher what was on each track. Some track labels can include kick, snare, bass, keys, synth, acoustic guitar, vocals, etc.

5. Headphones: If you are using headphones, go ahead and plug them in and make sure they are all receiving audio before the musicians arrive. It is also possible to ask the musicians what sort of headphone mix they would like *before* they arrive. This enables you to build a very rough headphone mix ready for when the artists arrive, freeing up lots of possible recording time.

6. Tune the drums: If there is a drummer and he or she plans on playing his or her own kit, have them arrive earlier than everyone else and make sure the drums are in tune. If you or the drummer doesn't know how to tune the drums, the drummer should take his or her drums to a local music store or drum shop and get the kit reheaded and tuned. It will be some of the best money the band can spend on the session! In tune drums rule!

▲ TIP

Troubleshooting a problem is part of an audio engineer's job. Showing up early, clearly labeling gear and tracks before the session, and making sure you are getting signal are some of the things that can help you with potential troubleshooting issues. Again, this is much easier to do alone, rather than with the band looking over your shoulder. Make sure everything is working before they show up and you will start off the session with a lot less stress.

Basic or Rhythm Tracks

Basics or rhythm tracks involve recording the "bed" of a song. This stage revolves around what the song will be built on and occurs after you have set everything up. Think of basic tracks as the initial stage of building a house. You must have a solid foundation to build a structurally sound house. At this stage, the song needs to be "in time" and "in tune."

Basic track details are as follows:

1. Establish time and tune: With songwriters you may record a *click* first to establish tempo and the length of the song and then an instrument track such as acoustic guitar or piano to establish the tune. If you are recording dance- or rap music, it is common to start with the beat and the hook or melody. This establishes the time and the tune. Songs are often created in programs such as Reason, Ableton LIVE, Acid, Reaktor, or GarageBand beforehand.

2. Scratch track: A scratch track is a track that you aren't expecting to keep. Its purpose is to provide a guide for the other musicians, so they know where they are in the song. Sometimes, it is also requested because the musicians are used to cueing off that particular instrument and they believe they will perform the song better with a scratch track. With a full band you are likely to record at least drums during this stage. If there are no drums, then the bass, rhythm guitar, and maybe a scratch vocal are recorded.

3. Multiple takes: If there are multiple takes, make sure to clearly mark which take is to be used. You would hate to build the song using the wrong take! Maintain track sheets and make sure your templates reflect the correct instrumentation.

4. Correcting rhythm tracks: It is common to fix bass mistakes or other rhythm tracks during this stage to make the foundation as solid as possible before overdubbing.

▲ TIP

Avoid using headphones during basic tracking if possible. Getting separate headphone mixes for each musician can take a lot of extra time. If you are working with a band, have the band rehearse the songs without vocals. Then when you record, the band can perform the songs all together without headphones. This is a great way to get that initial energy and avoid headphone mixes.

Overdubs

Overdubs take place after a solid basic track has been recorded. This is the stage where you begin to layer a song with additional instruments not played during the initial basic tracking. Make sure if the person overdubbing is monitoring with headphones that they have a good mix. Some people love to hear their instrument blasted in their mix, whereas other people want their instrument turned down low. Don't be afraid to consult the person overdubbing about what they would like to hear in their headphone mix. Really loud headphones can mess with pitch, so be careful not to blast them!

You can also do overdubs in the control room and the musician can listen over the monitors. See Figure 9.4. I prefer this method to headphones, because it makes for easier communication between you and the musician. Some people really like overdubbing this way, whereas others prefer to be in a separate room where they can get lost in the music without the pressure of being in the control room with the engineer. See Figure 9.5.

Overdub details are as follows:

1. Fix anything that needs fixing. Bad rhythm guitar track? Bad tone? Wrong bass note? This may be your last chance to make it better before you mix the song.

FIGURE 9.4

FIGURE 9.5

2. Solos: Now that the foundation is recorded it is time to spice it up! Solos are generally cut during this stage. Lead guitar players love this stage!
3. Vocal performances: Most lead vocal performances are overdubbed. A handful of singers may track their vocals during the basic stage to get a live or spontaneous feel. However, it is more typical for the singer to overdub the vocals so they can get it just right. Backing vocals are also overdubbed.
4. Doubling: Some instruments may be doubled, tripled, or even quadrupled during overdubs. You can thicken tracks and make them more interesting during this stage.
5. Additional instrumentation: Need strings? Horns? Slide guitar? Or other unique instrumentation? These additional layers are usually overdubbed.
6. Additional percussion: The time is now to add that shaker, tambourine, or other additional percussion.
7. Editing and clean up: It is common to edit throughout the recording process. Clean up unwanted tracks and erase bad takes to avoid confusion down the road.
8. In essence, overdubs, or overdubbing, are anything you record after the initial basic/rhythm tracks. Overdubs are the layers that add different colors and textures to the recording. Often one musician at a time will overdub. Some musicians will prefer monitoring with headphones in another room, whereas others may prefer to listen to the control room monitors and perform in the control room. Most instrumentation is generally completed during the overdubbing process. However, last-minute overdubs are often times performed during the mixing stage.

Rough Mixes

At some point, before, during, or after overdubbing, you may make some rough mixes of what you have recorded so far and hand them out to the musicians. Think of rough mixes as a quick sketch. Rough mixes can be used by the artist(s) to evaluate their performance(s). These mixes will give the band a grasp of what they have done and what they could possibly fix or add to the music. Rough mixes can also be very handy to other musicians who may be overdubbing at a later date. Rough mixes often end up in the hands of horn players, string players, a soloist, and backup singers. These rough mixes will provide them with a copy of the material they are going to overdub to at some point.

▲ TIP

Warning! You may want to point out to the band not to spend too much time listening to these rough mixes. Over listening to rough mixes is an affliction known as "demoitis" and has some real side effects. Occasionally, people will get so used to these mixes that no matter how great you mix later it will never surpass those first mixes. Keep in mind that rough mixes usually don't have the FX and mixing details that will occur when the song is finally mixed. I always tell my clients that they are called "rough" mixes for a reason.

Mixing

As I discussed in Chapter 6, mixing involves the blending and equalization of all the tracks to make a stereo, two-track, Left/Right mix of the material that you recorded. The exception would be for a mix for 5.1 or other surround sound formats that requires more than two tracks. I will usually perform any edits to the recorded material before I start mixing. However, you may perform additional edits during this stage.

Mixing tasks (please note, this is not an all-inclusive list) are as follows:

1. Balance of instruments: Balancing the volume between the recorded tracks is the first and most important step performed during mixing.
2. Compression or limiting: Controlling the dynamic range helps sounds sit more consistently in the mix while squashing any sudden peaks.
3. Panning: Panning helps the user separate like instruments and open up the complete 3D reference ball.
4. Equalization: Start with subtractive equalization then boost what is missing.
5. Additional editing: This is the engineer's last chance to clean up and consolidate the recorded tracks.
6. FX: Effects are applied at this stage for additional depth and texture. They can also be used to separate or glue instruments together.
7. Automation: Typically the last stage of mixing. Automation allows the user to program volume changes and other moves to later be recalled and performed automatically.

Creating a mix details (this is only one of many methods):

1. Get a quick balance between all the tracks.
2. Roll off bass on instruments or voices that shouldn't be sitting in the bass range. This could possibly include guitar, snare, some vocals (exception would be Barry White), tambourine, shaker, etc. This is a good way to clear up any unwanted mud in the mix and generally makes a mix much clearer.
3. Do the same thing to the high end-roll off any unwanted highs, likely on bass, kick, and any other low-end sounds that don't need the extreme highs.
4. Pan similar instruments to opposite sides along with any stereo mics. Try to separate instruments with panning before equalizing. If you have two electric guitars, try panning them to opposite sides, say 9 and 3 o'clock or hard left and hard right. Effective panning can open up a mix and give it more depth and dimension.
5. Although you will likely EQ instruments until the mix is done, you can usually start applying some compression before EQing a bunch of tracks. Compression will glue like instruments together. Compression also makes instruments more consistent. The kick drum, bass guitar, and vocals are often compressed. Of course, if compression is needed on other tracks, you could apply it to them as well.
6. Don't focus your attention on how an instrument sounds alone, unless, of course, it is the only instrument being heard. Instead, focus on how

the instrument sounds in the overall mix. Often, novices will EQ the ring out of a snare while soloing only to find that when the music is added back to the mix the snare is lost. Maybe the snare needs that ring to cut through the mix. I typically only solo instruments when I am searching for an unwanted sound, like a mic being bumped, distortion, or another extraneous noise.

7. Always start with subtractive EQing and then add what you are missing. (If you forgot what subtractive EQing is, refer back to Chapter 3.)

8. Don't wait until the end to introduce the vocals in the mix; you may not have enough room left by the time you get to them.

9. The lead vocal is typically the focus of the mix. Take extra care with the vocals. Are they loud enough? Can you hear all the lyrics? If you or the band can't agree where the vocals should sit, ask someone who hasn't heard the song before. I have worked with singers who insist their vocals are too loud and they can hear every word clearly. Of course, they would understand the lyrics at any volume because they likely wrote the lyrics. A neutral party is never a bad idea to include, but don't end up with too many cooks in the kitchen.

10. Once you have good balance, spatial positioning, EQ, and decent compression, add FX such as reverb, delay, and chorus. Like compression, reverb and other effects can help separate instruments and can also be used to glue things together. For instance, if you used a small plate reverb on the vocal and a medium hall on the drums, this could create some front and back separation. The vocals would appear more upfront because of the smaller space, and the drums would appear farther back in the mix because of the larger space. If you used just the medium hall on everything, there would be less separation between the sounds, and they would sound like they were all recorded in the same space, appearing cohesive. It just depends on what you are trying to do.

11. Let's say you want a particular instrument turned up during a certain section and then have it return to its original volume. This is something you would do during the automation phase of the mix. A distinct advantage of mixing digitally is that all your moves can easily be recorded and recalled at any time. There are recording consoles that have automation where the board can remember mutes, volume changes, and some can even be programmed to remember dynamic and EQ changes.

12. Start with mutes, especially at the beginning and end of the song. I will likely mute or erase any tracks that are not being played at a given time to make sure I don't hear mystery sounds later.

13. After performing mutes in automation, you can move on to volume changes.

14. Make sure that you aren't overloading the stereo mix levels if you are mixing a digital two-track or bouncing to stereo. Don't worry about getting the levels as loud as possible. Maximizing volume isn't necessary if you are recording 24 bit. Don't go over ZERO in the digital world! Every

engineer has his or her own preference; I set my stereo two-track levels around −6 dBFS. Some engineers prefer more room for transient spikes and will set the level average around −12 dBFS. If, by some chance, you happen to mix down to an analog two-track, you will likely push the levels above zero to get the best signal to noise ratio.

15. Monitor on several systems. Don't just use headphones! You want your mixes to sound good on several systems. Bass can be hard to properly evaluate with headphones. Headphones can be deceiving but can be helpful for checking instrument placement and listening for specific details. I have a cheap stereo jam box that I use to monitor. Think about setting up a monitor system that has both accurate speakers and specialty speakers such as Auratone or Avatone mix cubes.

16. Make sure you note which mix or mixes that you plan on using as the final mix. You would hate to give the band or the manufacturer the wrong take.

▲ TIP

Compare your mix to mixes with which you are familiar. Although a mastered CD will sound louder, you can still compare the different frequency ranges and make sure your mix has enough of each frequency range.

Oh, and did I mention listen? Make sure you take time to listen to how each frequency range sounds. Throughout the mixing process you should listen not only to an individual instrument, but how that instrument relates to the overall picture. Don't get obsessed with one particular sound, but instead get obsessed with the sounds as a whole. Many inexperienced engineers will key on one track, say the kick drum. The kick will sound awesome and everything else sounds weak and poorly represented. After all, mixing is about relationships and how all the different sounds work together.

Keep in mind that this is only a guideline of one way to mix a song. Other engineers may go about the mixing process very differently. Just like artists, engineers can paint a picture many ways. Mixing takes a lot of practice. The way you get better at mixing is to mix as much as possible. Remember, it is how everything sounds together. Don't spend too much time in solo mode!

Mastering

Mastering is the final stage in the recording process. During this stage, your final mixes or *stems* are assembled together to make a stereo master from your recording sessions. Many people aren't aware of this stage but this is what can make a demo sound like a finished product. Professional recordings are mastered. Some engineers do both the mixing and mastering of a recording. However, many engineers prefer to stay clear of mastering a project they have personally recorded and mixed, recommending the final mixes be handled by an objective mastering engineer. Mastering prepares your mixes to be sent off for *duplication* or *replication*.

Mastering details are as follows:

1. Equal levels: Mastering will ensure that the levels between all the songs are relatively equal. Overall levels will be consistent if your music is put on shuffle or if you skipped from one song to another. Nobody wants to constantly adjust the volume between tracks.

2. Equal tone: Along the same lines, mastering also ensures that the EQ between tracks is the same. Just like with volume, you wouldn't want to constantly adjust the tone between tracks. Mastering will keep the listeners from constantly adjusting their tone knobs. You wouldn't want the listener to have to add bass on one song, then turn around and cut the bass on the next song. Not cool!

3. Compression: Make sure you leave plenty of headroom with your mixes. This will allow the mastering engineer to perform his or her job to the fullest. There isn't much the mastering engineer can do if you give them a stereo mix that looks like a solid block from top to bottom or is maxed out at 0 dBFS.

4. Song order: A big part of mastering is sequencing the recording. The mastering engineer does not decide the order. Song sequences are usually provided to the mastering engineer by the band, record label, or producer of the project.

5. Editing: What if you wanted to eliminate an entire section and paste the remaining parts together after the song has been mixed? You could even pick the best chorus, copy it, and insert it over the other choruses that weren't as good. This is something that could be done in mastering. Most mastering is done digitally, a nonlinear editing format; therefore, the possibilities are pretty endless.

6. Insert space between tracks: Should the songs be back to back with little or no time between songs? Or should the space vary from song to song? Maybe more empty space between a fast and a slow song? Or maybe a nice cross-fade between the two tracks? Ultimately, you will have to listen a few times to determine the spacing and timing between tracks. What happens here can directly impact on how an album flows from start to finish.

7. CD Text: Would you like the artist's name to appear on screen when the CD is played? Any CD Text or other data can be added during mastering. As a result, when you load a CD into a computer, car, and some CD players, the artist name, album title, and song name will appear along with any other information you would like included. CD Text is an extension of the *Red Book* specification for audio CDs that allows for storage of additional text information on a standard audio CD.

8. Index and time information: Your master will also include any indexing and time information that is needed by the company that is going to duplicate or replicate the CD, DVD, or vinyl record. Some duplication companies use a DDP image (Disc Description Protocol) format for specifying content with CDs, DVDs, and optical disc.

9. ISRC codes: Other information such as *ISRC* codes would be embedded at this time. ISRCs are increasingly being used by major download sites, digital distribution companies, and collecting societies as a tool to manage digital repertoire and to track commerce.

10. Master copies: Finally, you will receive at least one Master copy of your material and likely a Safety Copy. You will probably receive one to four additional copies to listen to while you wait for an official copy with artwork and all. You may request a test pressing of the record if you are pressing to vinyl.

Mastering fees are typically separate from engineering fees. Some mastering engineers charge per song, per project, or just an hourly rate. Other fees can include per Master CD and for additional copies of the Master. A few engineers may also charge you extra if there are multiple or difficult edits. I make sure there is always at least some minimum mastering work on my recorded projects. Mastering can help glue the songs together and make the overall tone more consistent. With the right mastering your mixes will be taken to the next level!

FOUR QUESTIONS TO ASK BEFORE YOU RECORD A BAND

Now that you are familiar with the seven stages of the recording process, let's discuss how to handle a potential client. These questions are important, since you are likely to be self-employed as a recording or live sound engineer.

Before a recording session, you should ask the band or artist the following four questions:

1. What is your budget? This will determine how much time to spend on each stage. The budget typically determines when the project is over. It will also determine where we record.
2. What is your instrumentation? This helps with the setup stage but also helps decide if the budget fits the expectations. For instance, it would probably take less time and be an easier session to record just two acoustic guitars and a lead vocal versus recording drums, bass, electric guitar, keyboard, lead vocals, backing vocals, and a trumpet player.
3. What is the purpose and length of recording? Is the recording for booking shows, a press kit, label release, or movie soundtrack? A demo to get shows is much different than an actual label release. You will also need to know how many songs and the length of the songs to make sure enough studio time is booked to complete what the artist would like to accomplish.
4. What are your expectations? Is the person or persons you are recording expectations higher than their budget? Often people want a Porsche for the price of a Pinto.

As you gain more experience, determining a budget becomes easier and more predictable. If you can't meet an artist's expectation, don't take the work. It won't be good for either side. People generally want more for their money (would you blame them?), so it will be up to you to determine if it will be worth the price you agree on.

In Chapter 12, I am going to talk about internships. One of the most valuable tools I learned from my internship was how a recording session progress. Hopefully, this chapter has made you aware of how a recording session flows from start to finish. Keep in mind that no two recording sessions are alike. That's what keeps it interesting. Experience counts for a lot when recording, and in most things, for that matter. Every new recording experience will help you become a better audio engineer and give you more insight on how a recording session can be run smoothly.

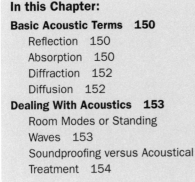

CHAPTER 10

Basic Acoustics...How to Make Your Recording Space Sound Better

149

If you get into music production and decide to have your own home or project studio, you will need to learn some basics about acoustics. Acoustics is the science of sound and can help determine the quality of sound transmitted in a room or space. Understanding sonic characteristics of a room will result in better decisions when recording or mixing.

Different surfaces and materials can affect the acoustics in a room. Too much of one material applied to a wall or surface is generally destructive to a sound. For instance, if a room is carpeted from floor to ceiling, the room will sound dead and dark. Carpet does an exceptional job absorbing higher frequencies and minimizing hard reflections. On the other hand, it does little to deter the long low-end waves. Whether it is carpet, concrete, tile, or any other material, too much of one type of material is rarely a good thing. Professional recording environments are typically constructed with many types of surfaces and materials. They may have wood floors with throw carpets, a brick or stone wall, diffusers, a few concave or convex walls, and a variety of other acoustic treatments. Unlike most bedrooms or home offices, where many project studios are located, professional recording spaces are rarely perfect squares with low ceilings. This is because parallel walls and surfaces create standing waves, or room modes, and can affect the frequency response of your room. Many professional recording studios are acoustically designed, especially the control room.

▲ TIP

Don't worry if you are unable to hire an Acoustician to setup your studio. As with speakers, you are not required to have the perfect pair, or in this case the perfect space; knowing your space will allow you to consistently predict the way a mix will translate through other listeners' audio systems.

We just discussed that carpet absorbs only a limited amount of the lower frequency range. If you completely carpet a room with the intention of soundproofing, the carpet will do very little to stop the long, destructive bass waves from traveling outside. If the neighbors are offended by loud music, carpet will not solve the problem. Later in this chapter, the distinction will be made between soundproofing and acoustic treatment.

This chapter will include only basic acoustic terms and concepts, as the science of acoustics is too complex to be covered in a beginner's guide to audio. This chapter also provides a do-it-yourself section to help you learn how to make your room sound better by building acoustic treatments such as gobos, diffusers, and bass traps.

As illustrated in Chapter 2, sound is divided into three successively occurring categories. Sound is made up of the direct path, its early reflections, and the reverberant field. With a basic understanding of acoustics, it will be much easier to manage these three elements of sound.

BASIC ACOUSTIC TERMS
Reflection

When sound strikes a wall, ceiling, or any object, the sound is reflected at an angle equal to, and opposite of, its initial angle of incidence. The surface material determines the amount of energy reflected. Sound and light have similar reflective paths.

▲ TIP

You may want to step outside for this one. Try throwing a tennis ball or other bouncy ball at the floor or ground near a wall. Notice how the ball rebounds, or reflects, off multiple surfaces. This is similar to how sound would reflect off the same wall.

Absorption

Absorption is the process in which acoustic energy is reduced when a sound wave passes through a medium or strikes a surface. Absorption is the inverse

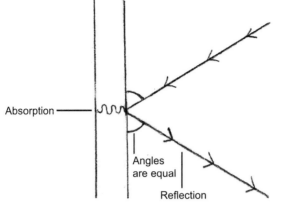

Absorption

Angles
are equal

Reflection

FIGURE 10.1

of reflection. The amount of energy absorbed compared to the amount reflected is expressed as a ratio known as the *absorption coefficient.* Different materials affect different frequencies (see chart below). If a material's absorption coefficient is 0.25, it absorbs 25% and reflects 75% of the sound. An absorption coefficient of 1 means that 100% of the energy striking the surface will be absorbed and none reflected. An absorption coefficient of 0 means none of the sound is being absorbed by the surface and 100% is being reflected.

See how sound is affected by these different materials.

Frequency	125 Hz	250 Hz	500 Hz	1 kHz	2 kHz	4 kHz
Concrete/ unpainted	0.36	0.44	0.31	0.29	0.39	0.25
Concrete/painted	0.10	0.05	0.06	0.07	0.09	0.08
Light drapes	0.03	0.04	0.11	0.17	0.24	0.35
Heavy drapes	0.14	0.35	0.55	0.72	0.70	0.65
Carpet	0.02	0.06	0.14	0.37	0.60	0.65
Plywood	0.28	0.22	0.17	0.09	0.10	0.11
Glass windows	0.10	0.05	0.04	0.03	0.03	0.03
Marble/tile	0.01	0.01	0.01	0.01	0.02	0.02

Notice that plywood and unpainted concrete absorb lower frequencies the best, whereas carpet and heavy drapes are most suitable for absorbing higher frequencies. Ultimately, the goal is to combine materials so your room has a flatter frequency response.

Diffraction

The ability of sound to bend around an object and reform itself is known as diffraction. Bass frequencies are great at this! Have you ever heard bass from a car trunk thumping through the neighborhood? The reason why only the bass range is heard is that low frequencies can easily escape the confines of the car or trunk. The lack of higher frequencies is due to the fact that objects such as buildings, trunks, trees, etc., easily impede more directional, higher frequencies.

Diffusion

Diffusion is a way to control reflections and standing waves. A diffuser can be used to break up standing waves by reflecting sound at a wider angle than its initial angle of incidence.

As previously mentioned, different surfaces and materials will reflect and absorb sounds differently. Figure 10.2 shows how sound reflects off a 90° angle concave surface, convex surface, and straight wall.

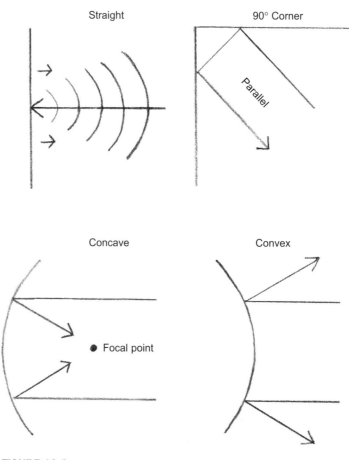

FIGURE 10.2

DEALING WITH ACOUSTICS
Room Modes or Standing Waves

Standing wave

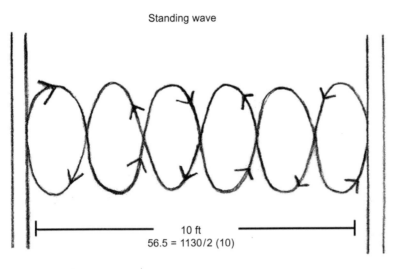

10 ft
56.5 = 1130/2 (10)

FIGURE 10.3

Room modes, or standing waves, are created by parallel surfaces. Acoustical problems may arise from standing waves such as a level boost or cut off frequencies, or a creation of resonant frequencies in a room.

Have you ever sung in a bathroom or other reflective space and noticed how a particular frequency resonates in the room? Standing waves create this resonance. Here is how to calculate a room's standing waves.

Frequency = 1130 (speed of sound or velocity) divided by the length × 2

For instance, with a 10' space between two walls, the equation will be:

$$1130/2(10') = 56.5\,\text{Hz}$$

This would be the first resonating frequency in the center of the room. There will be many room modes between two walls as the phenomenon will repeat itself at multiples of the first frequency: 2*f*, 3*f*, etc. Therefore, the second room mode would be 56.5 × 2 = 113 Hz, the third would be 56.5 × 3 = 169.5 Hz, and so forth.

 TIP

How to deal with standing waves:

Trapping bass is one way to lessen standing waves in a room. Diffusers and other acoustical treatments may also be helpful.

Soundproofing versus Acoustical Treatment

It is important to make a distinction between soundproofing a room and treating the room to sound better. Soundproofing a room is much more difficult than acoustically treating a room.

Soundproofing involves completely isolating sound so it does not escape the room. It is much more difficult and expensive to soundproof after a room has been built versus preplanning and building the room from the ground up. Air space is a great way to soundproof. Building a room within a room and leaving a small space between the new walls and the original wall will be very effective in isolating sound. Believe it or not, a little air space between layers is good for isolation because it helps slow the sound's momentum.

Acoustical treatment involves making a room sound better. Generally, this can be achieved much more cheaply than completely isolating a room. Acoustical treatments include bass traps, diffusers, Auralex, or other wall foams. These treatments may involve minor construction, but generally not to the degree of soundproofing a studio. There are many do-it-yourself (DIY) methods and resources. Here are a few of those DIY projects.

DIY PROJECTS
How to Make Your Room Sound Better

First off, identify your room's acoustical qualities. Use the standing-wave/room-mode calculation previously discussed to identify potential sound quality issues.

There are also many software programs that offer tools to help identify acoustical qualities. A measurement microphone can be used with a variety of software programs or spectrum analyzers to help identify your room's anomalies.

Here are a few RTA software programs available:

Faber acoustical, www.faberacoustical.com – for Mac and iPhone

True Audio, True RTA, www.trueaudio.com – for PC

ARTA, Audio Measurement and Analysis, www.fesb.hr/~mateljan/arta/index.htm – Software/Shareware

🔺 TIP

As mentioned previously, meters are only tools. Always trust your ears first. The meters can help verify what you are or are not hearing.

How to make your bedroom sound better:

1. Use smaller speakers.
2. Set up symmetrically.
3. Use a bed or a couch to help with bass control and to tighten up the room sound.

4. Don't set up your mix station and speakers in the corner or with your back too close to the wall.
5. If your room is tiled or concrete, place a carpet or two on the floor to dampen the reflections and echo.
6. Don't cover all your walls with the same material or treatment. This may end up causing another set of problems… too much of one material is detrimental to good sound quality.
7. Cover your windows with heavy curtains to keep sound from escaping.
8. If needed, build bass traps to control the bass. See the DIY project below.
9. Try stacking empty boxes in the corner of the room. Place the open side against the wall and fill them with sheets, dirty laundry, blankets, pillows, or similar material to help trap and control bass.
10. If needed, build a diffuser to help break up and disperse sound for a more accurate sounding space. Again, see the DIY project below.
11. If needed, treat walls and ceilings with the necessary acoustical materials.
12. If possible, remove appliances or noisy electronics from the room.

There are many acoustical treatments you can build on your own. These treatments will help you control bass, midrange, and treble within a recording space, allowing you to effectively manage a room's imperfections. Three common, easy-to-build contraptions are gobos, diffusers, and bass traps. I suggest treating your space over time and tweaking it as you go. The following projects should help you affordably address at least a few acoustical issues.

How to Build a Gobo

7   How to Build a Gobo

A go-between, also known as a gobo, is a great mobile way to control sound. It can be used to help isolate an instrument or lessen the reflections and reverberation of a space. I often place a gobo in front of an amplifier or kick drum to isolate and absorb the sound. Gobos are also effective if placed around a drum kit, especially if they are being played in a "live" space such as a kitchen or on a tiled floor. This will tighten up the sound and minimize reflections.

In order to build a gobo, you will need the following items:

1. R11/R13 wall insulation or Owens Corning 703
2. Jute or polyester material
3. 1"–2.5" wood screws
4. Stapler or staple gun
5. Utility knife
6. Scissors
7. Philips screw driver, but preferably a drill
8. Saw
9. 1×2, 2×2, 1×4 wood
10. Two small washers

How to build a gobo:

1. Build the frame to the desired size.
2. Cover one side with the jute, burlap, or other material by stapling to the frame.
3. Put the insulation inside.
4. Enclose insulation with material and staple to frame.
5. Finish off with additional framing/trim if desired.
6. Attach feet with a single screw and washer. Don't tighten the feet too much, allowing the feet to swivel.

FIGURE 10.4

You can put plywood or masonite on one side for a more versatile gobo. One side will be hard and reflective and the other soft and absorptive. The plywood or masonite will absorb some of the lower frequencies.

⚠ TIP

You can also use a mic stand with a boom arm and acoustic blanket for a quick, mobile gobo. This is a great way to quiet down a drum kit or tighten up an amp or vocal recording. Fold the mic stand into a "T" and drape the blanket over it. Easy and fast to set up and to tear down. Many hardware stores and moving companies sell quilted packing or acoustic blankets. They are also available at companies that carry acoustical treatments.

How to Build a Diffuser

A diffuser is a great way to control reflections in a room and make your room sound flatter. If your room is on the "dead" side, diffusers may be more suitable than adding absorptive sound control. Diffusers are often placed on the back wall behind the engineer's head. They can also be placed in the studio to help tighten up a room, reduce *flutter echo*, and control reflections. Here's a good way to make a diffuser using a proven formula. This diffuser is designed to be effective up to about 600 Hz.

In order to build a classic quadratic diffuser based on the BBC RD report from 1995 (Figure 10.5), you will need the following items:

1. 1×1, 1×2, 2×2 wood of your choice, 8' length. Quantity dependent on size of diffuser
2. Heavy duty wood glue
3. 12" × 12" or 24" × 24" ¼" masonite or other wood
4. Heavy picture hangers or screws for mounting

How to build a diffuser:

1. Cut the following pieces to the appropriate length listed below. You will have a total of 131 pieces with 13 blank spots.
   - 1"−38
   - 2"−38
   - 3"−40
   - 4"−15
2. Arrange the precut pieces of 1×1 or 2×2 wood and glue them on the ¼" 12" × 12" or 24" × 24" masonite or other wood base in the following manner:

0	3	4	1	2	3	3	1	4	2	3	3
3	0	1	4	2	1	1	3	3	2	1	1
3	1	1	3	1	3	2	2	1	0	2	2
2	2	2	2	0	4	3	2	3	2	1	1
3	3	1	1	3	1	1	3	4	3	1	3
2	3	2	1	2	0	3	2	4	2	1	0
2	3	2	1	3	1	2	2	3	1	3	4
2	0	2	4	4	0	1	2	1	4	2	2
3	4	1	0	1	3	3	1	0	2	3	3
1	3	3	1	2	4	1	2	0	1	3	1
2	1	2	3	1	3	3	2	4	2	3	4
2	4	2	3	3	1	1	2	0	3	1	0

3. Hang and enjoy.

FIGURE 10.5

How to Build a Random Diffuser

In order to build a random diffuser, you will need the following items:

1. Scrap wood
2. Heavy-duty wood glue
3. Any size piece of plywood for use as a base
4. A handful of wood screws

How to build a random diffuser:

1. Use your imagination and arrange the scrap wood as you wish on the plywood. Although a random diffuser may not be to a specific formula, it will help break up and disperse sound effectively.
2. Glue scrap wood to base wood.
3. Hang and enjoy.

FIGURE 10.6

How to Build a Bass Trap

Bass traps are effective in capturing and controlling low end and are often placed in room corners because this is one place where bass lingers. One of the easiest bass traps to build is known as the "super chunk." The materials are fairly inexpensive and it does the trick.

In order to build a bass trap, you will need the following items:

1. 12 panels of 24" × 48" × 2" mineral wool or rigid fiberglass board or an 8' ceiling
2. 1 × 4 or 2 × 4 wood for framing
3. Jute or polyester
4. Stapler or staple gun
5. Drill
6. Razor blade, kitchen knife, or other cutting edge
7. Gloves for cutting insulation
8. Glue for insulation

How to build a bass trap:

1. Cut the panels in half, this should give you plenty of 24" × 24" × 2" squares.
2. Next, cut these squares into triangles.
3. Glue and stack triangles from floor to ceiling in corner of control room or studio space.
4. Make sure stacked triangles are pushed up against the wall.
5. Build a rectangular frame out of the 1 × 4s to cover the stacked triangles.
6. Wrap the frame with material, preferably jute or polyester, and staple it to the inside of the frame.
7. Screw frame into corner hiding the stacked triangles.

Learn before you build. Listen to your mixes and get to know how your room sounds before you get excited and staple egg crates all over your walls. Take your time and get it right. Read more about acoustics, get to know your recording environment, and tweak your room as needed.

If you are not the DIY type or just don't have the time, you can purchase gobos, diffusers, bass traps, broad band absorbers, and other acoustic treatments at a variety of reputable companies. Some of these companies are listed below.

Additional Links and Resources

http://www.acoustimac.com/index.php/ecoustimac-eco-friendly.html

www.auralex.com

www.realtraps.com

www.tubetrap.com

www.atsacoustics.com

www.readyacoustics.com

CHAPTER 11

The History of Audio: It Helps to Know Where You Came From

Music no longer has to be stored on reels of tape. Digital audio has taken the worry out of having to physically deliver or ship a master tape of the band's hit songs at the mercy of the carrier. Even into the mid '90s, it was common for engineers to travel with their master tapes, or for the masters to be mailed. There was always a chance these masters could be destroyed by the x-ray machine or damaged during transit. Nowadays, a master session is easily transferred over the Internet and mixes can easily be shared online through a variety of methods. This is why it is important to understand different file formats, compression methods, and how audio files such as MP3s, AIFFs, and WAVs affect audio quality.

To fully understand digital audio technology, you need to understand how, when, and why we ended up digitizing audio. For over a century, analog was the only way we recorded and reproduced music. Digital technology opened up recording to the masses. Prior to the introduction of digital technology, most musicians frequented large, professional studios to record their music, because there was no other option. These days, many musicians and engineers have the option of using professional home or project studios to record music.

Many extraordinary people contributed to this modern era of recording. Legendary engineers such as Les Paul, Tom Dowd, and Bill Putnam shaped the audio world as we know it today. Numerous audio advances occurred around World War II as the world at war sought out new communication technologies. Until the 1980s, music was recorded primarily to an analog medium such as lacquer or tape. Although the technology to record music using a computer had been around for more than fifty years, it wasn't until twenty years ago that it became commonplace. Currently, many people combine the best of both worlds, recording with both analog and digital equipment.

AUDIO HISTORY
A Brief History of Recording

Sound was initially recorded to a single track with a single transducer. Early on, it was extremely important to place the one and only transducer in the perfect spot to pick up all the musicians and get a good balance. This miking technique applies even today when there is no other option. Eventually, music would be recorded using many mics and recording to one or two tracks. In the '50s, people like Les Paul started using eight tracks with more mics and began to regularly layer (overdub) tracks. This was really the beginning of modern recording as we know it. Now, music production is much more flexible, allowing engineers to record limitless tracks in their own home with an array of affordable equipment.

Some notable moments in sound and recording:

- The first method of recording and playing back of sound is credited to Thomas Edison with his invention of the phonograph in 1877.
- The first flat, double-sided disc, known as a record, was invented by Emile Berliner in 1887. That same year, Berliner also patented the gramophone.
- The first 10-inch 78rpm gramophone record was introduced in the early 1900s. The 10" offered around 4 min recording time per side.
- Stereo recording was patented in 1933 by Alan Blumlein with EMI.
- The LP (long-playing) record was invented by Columbia Records in 1948. LPs offered around 30 min recording time per side.
- The first multi-track recording was developed in Germany in the 1940s; however, the first commercial multi-track recording is credited to Les Paul in 1955.
- The first 7-inch 45rpm record was introduced in North America in 1949. The 7" offered around 5 min recording time per side.
- Ampex built the 8-track multi-track recorder for Les Paul in 1955.
- Endless loop tape existed decades before the classic eight track player and cartridge were invented by Bill Lear in 1963.
- Although Elisha Gray invented the first electronic synthesizer in 1876, Robert Moog introduced the world to the first commercially available synthesizer in 1964.
- Before transistors, vacuum tubes were the main components in electronics. Today, the transistor is the main component in modern electronics. The transistor has contributed to things becoming smaller, cheaper, and lasting longer.

Although the original transistor patent was filed in the mid '20s, transistors were not commonly used to replace vacuum tubes until the '60s and '70s.

- Cassettes were invented by the Phillips Company in 1962, but did not become popular until the late '70s. Cassettes took up less physical space than vinyl records.
- MIDI (Musical Instrument Digital Interface) was developed in the early '80s, which enabled computers and electronic musical devices to communicate with one another (MIDI is discussed later in this chapter).
- James Russell came up with the basic idea for the Compact Disc (CD) in 1965. Sony Philips further developed the idea and released the first disc in 1979. CDs were introduced to the public in 1982. Just as cassettes replaced vinyl, CDs brought the cassette era to an end.
- Pro Tools, originally released as "Sound Designer," was invented in 1984 by two Berkeley students, Evan Brooks and Peter Gotvcher. Pro Tools was introduced to the public in 1991 and featured four tracks costing $6000.
- The ADAT-acronym for Alesis Digital Audio Tape. The ADAT recorder helped usher in the home project studio by making a compact and affordable multi-track tape machine. It was introduced by Alesis in 1992.
- Although the initial idea for the MP3 was developed in the early '70s to compress audio information, the MP3 wasn't adapted until about 1997. An MP3 requires less physical storage space than its many predecessors: the vinyl record, cassette, and CD. In fact, it takes almost no physical storage space! Imagine storing 10,000 vinyl records as opposed to storing 10,000 MP3s.

Innovators Who Contributed to Modern Recording

As previously mentioned, Bill Putnam, Les Paul, and Tom Dowd are three of the early pioneers of modern recording. These people introduced the modern recording console, multi-track recording, reverb, tape delay, multiband EQ, the fader, and many other developments now taken for granted.

BILL PUTNAM

Often referred to as the "father of modern recording." His company, Universal Recording Electronics Industries (UREI), developed the classic UREI 1176LN compressor, a signal processor that is highly regarded to this day. Putnam is credited with the development of the modern recording console and contributed greatly to the post WWII commercial recording industry. Along with his good friend Les Paul, Bill helped develop stereophonic recording. He is also credited with developing the Echo send and using reverb in a new way. Not only was Bill Putnam a highly sought after engineer and producer, he was also a studio and record label owner, equipment designer, and musician. Bill Putnam started Universal Recording in Chicago in the '50s and recorded such artists as Duke Ellington, Count Basie, Hank Williams, Muddy Waters, and Frank Sinatra. He eventually moved his company to California and renamed it United Recording Corp. Bill passed away in 1989 and in 2000 he was awarded a Grammy for Technical Achievement for all his contributions to the recording industry.

LES PAUL (LESTER WILLIAM POLSFUSS)

You can't talk about early recording without mentioning Les Paul. Most people know of Les Paul because of his legendary guitar playing; but many engineers know of Les Paul for his contributions to the recording industry. Although not the first person to incorporate layering tracks, Les Paul is credited with multi-track recording as we know it. Les Paul had already been experimenting with overdubbed recordings on disc before analog tape. When he received an early Ampex Model 200, he modified the tape recorder by adding additional recording and playback heads, thus creating the world's first practical tape-based multi-track recording system. This eight-track was referred to as the "Sel-Sync-Octopus," later to be referred to as the "Octopus." Les Paul is also credited with the development of the solid body electric guitar. Many believe that this instrument helped launch rock n' roll.

TOM DOWD

A famous recording engineer and producer for Atlantic records. Tom worked on the Manhattan Project that developed the Atomic bomb before he started his extraordinary career in music production. Tom Dowd was involved in more hit records than George Martin and Phil Spector combined. He recorded Ray Charles, The Allman Brothers, Cream, Lynard Skynard, the Drifters, the Coasters, Aretha Franklin, J. Geils Band, Rod Stewart, The Eagles, and Sonny and Cher, to name just a few. He also captured jazz masterpieces by Charlie Parker, Charles Mingus, Thelonius Monk, Ornette Coleman, and John Coltrane. He came up with the idea of a vertical slider (fader) instead of the rotary type knob used at the time. Like Putnam and Paul, Dowd pushed "stereo" into the mainstream along with advancing multi-track recording. Tom Dowd was also one of the first engineers willing to make the bass line prevalent in modern recordings. Dowd was an incredible musician known for his remarkable people skills and received a well-deserved Grammy Trustees Award for his lifetime achievements in February 2002.

▲◉ TIP

Check out the DVD biography of Tom Dowd, "Tom Dowd, The Language of Music."

ANALOG AND DIGITAL AUDIO
What Is Analog?

Sound that is recorded and reproduced as voltage levels that continuously change over time is known as analog. Examples of analog are cassette tapes, vinyl records, or analog recorders. The up and down variations in pressure levels that sound creates are represented in the grooves of a record or the arrangement of magnetic particles on a tape. Analog gear can be purchased in the form of FX processors, amplifiers, keyboards, tape machines, compressors, audio production consoles, and other electronic components.

FIGURE 11.1

THE PROS AND CONS OF ANALOG

Pros

Tonal quality: Proponents of analog often describe it as fat, warm, and easy on the ears.

The soul: Many audiophiles and purists swear by analog and the way that it captures more of the soul of the music.

Transients: An analog recorder softens the transients. Unlike digital audio, analog doesn't reproduce and record transients as well. Many audio engineers enjoy this because cymbals and other edgy or piercing sounds are toned down and mellowed.

Tape compression: In the digital world, levels can't go over zero, but with analog it is quite often desired. When the level is pushed on an analog recorder, the signal is saturated on the tape, a desired effect for many.

Holds its value: Quality analog equipment typically holds its value and can easily be resold.

Classic: Analog tape provides a classic tone. Especially useful for certain styles of music with more of a roots or organic feel such as classic rock, jazz, folk, bluegrass, blues, surf, indie rock, and some country.

No Sampling: With digital audio, samples are taken of the music. Analog means it is analogous to sound. With analog you get all the pieces and not samples. Therefore, the music isn't relying on the listener to fill in the gaps between samples, especially with low-quality lossy files. Lossy files will be discussed in greater detail later in this chapter.

Cons

Editing: Although almost any edit can be done in the analog world that can be done in the digital world, editing is destructive and time consuming. With analog, the tape has to be cut physically and there is no "undo," there is only "do-over"!

Cumbersome: A typical analog recording setup requires more physical space than most digital setups.

Conversion: Most recordings end up digital anyway (unless you record all analog and press to vinyl), so why not just start with digital? Obviously, you can't download or upload music in an analog format. It has to be digitized at some point.

Tape costs: A reel of 2" tape needed for analog recording cost about $250. A standard 2500' reel of tape can record about 16–33 min of material, depending on the tape speed.

Sound quality: Digital enthusiasts describe analog as noisy and muddy. Many like the clarity that digital audio provides.

What Is Digital?

Unlike analog recording, digital recording is not continuous in that the samples of sound are taken and reconstructed to appear like a sine wave. Digital's main components are sample rate (related to frequency) and bit depth (related to amplitude). Digital audio technology at its most basic level is a means of encoding data through the use of the binary number system. Digital audio translates the alphabet, base 10 numbers, and other types of information into 1s and 0s, on/off voltage.

THE PROS AND CONS OF DIGITAL
Pros

Editing: The editing capabilities in the digital audio world are undoubtedly the high point. These editing capabilities are non-destructive. Sections of songs can easily be manipulated without damaging the quality of sound. There is an "undo" function.

Efficient: It is much easier for an engineer to flip from one song to another or from one recording to another during a session. With analog tape you may have to change tape reels or rewind the tape. With digital, a click of the mouse allows you to flip between recorded projects.

Transients: Digital audio reproduces transients accurately. This is one of the reasons why it is known for clarity.

No noise: Tape hiss and extraneous noise will most likely not be a problem in the digital world. You don't have to worry about recording a hot signal to cover the inherent noise associated with analog tape.

Compact: In the digital world, a full recording setup can literally exist in the palm of your hand. Analog equipment requires more physical space.

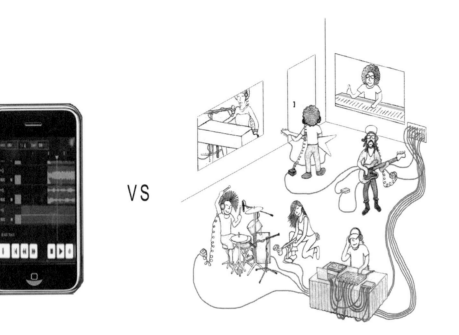

VS

FIGURE 11.2

Convenient: You can stream your favorite music from any country with online sites like SoundCloud (www.soundcloud.com). You can also store songs, mixes, and full playlists on many wireless phones. It would be very difficult to lug around your favorite 500 albums.

Storage capacity and costs: A hard drive or other digital storage device is cheaper and can store hundreds to thousands of hours of recorded material in one place.

Recall: You can recall any previously stored setting of a mix. If the band loves the final mix, but just wants the bass up a dB or two, you can easily open the last mix file and adjust. In the analog world, unless the console is completely automated, you have to make a fresh start. Even then, you have to document every signal processors' settings and any other settings that cannot be automatically recalled.

Demand: People demand audio digitized. Whether it is in the form of an MP3 or a streaming audio file, digital audio is the standard.

Cons

Tonal quality: Early on, digital audio was very harsh and although it has come a long way, there are engineers who still wish it sounded more like analog. Digital audio is often described as sounding thin and bright.

Value: Digital equipment is like buying a new car. Once you drive it off the lot, it loses value. It is not lucrative to re-sell outdated software or digital audio components.

Lossy files: Lossy files throw away "unnecessary" information to condense an audio file's size. Unfortunately, once you convert an audio file to an MP3, or other lossy format, it cannot be returned to its original audio quality.

It can crash: If audio files are not backed up and the DAW crashes, you could lose everything you recorded.

Digital Audio Terms

The following terms should help you increase your understanding of digital audio basics:

The Nyquist Theorem: States that the signal's sample rate must be at least two times greater than the highest desired frequency. For example, a sample rate of 44.1 kHz represents sound up to 22.05 kHz, and a sample rate of 48 kHz represents sound up to 24 kHz.

Sample or sampling rate: Factor in determining the frequency range in digital audio. The sample rate determines how many samples (pictures) of the audio signal are taken in a one-second period. A common sample rate of 44.1 kHz means that over forty-four thousand samples are taken per second. By cutting the sampling rate in half, we can determine the highest frequency that will be recorded and reproduced. For instance, a CD's standard sample rate is 44.1 kHz, which represents up to 22.05 kHz. Other common audio sampling rates are 48, 88.2, 96, and 192 kHz.

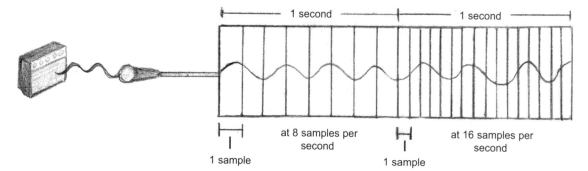

FIGURE 11.3

Quantization: Refers to the amplitude component of digital audio and determines how many steps or calculations are made. The more steps there are and the smaller the distance between each step, the smoother the representation. 8 bit = 256 steps, 16 bit = 65,536 steps, and 24 bit = 16,777,216 steps. Basically, the higher the bit resolution, the more the sound wave will appear as a sine wave.

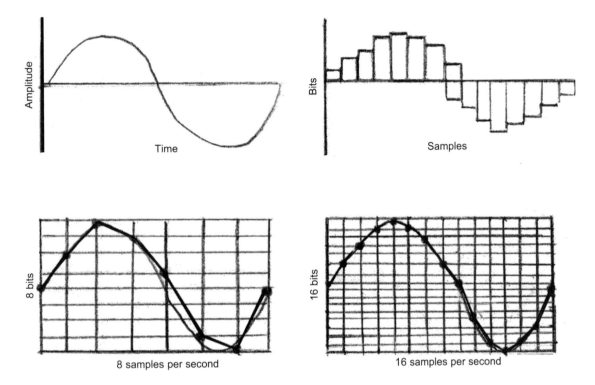

FIGURE 11.4

Bit depth: Determines the dynamic range. Common audio is 16- and 24-bit. Higher bit depths increase the resolution of the audio. Think of bit depth as the knob on a pair of binoculars that allows you to sharpen the image. Higher bit depth, in this case, makes a digital sample of square steps appear smoother and much more like an analog sine wave.

Bit rate: The rate at which digital audio is transmitted. It is expressed in bits per second, generally written as bps.

Normalize: A gain-related process in digital audio where the volume of the entire file is increased to a pre-determined standard. Unlike compression, normalizing a file does not change the dynamic relationship between the tracks. It keeps volume levels steady from song to song. This is useful to know if you are a DJ, creating mix tapes or trying to bring up low volumes in recordings.

Digital Audio Comparisons according to bit depth, sample rate, bit rate, and file sizes.

			Stereo Audio	
Bit Depth	Sample Rate	Bit Rate	File Size: 1 min	File Size: 10 min
MP3	128 k/bit	0.13 Mbit/sec	0.94 MB	9.4 MB
16	44.1 kHz	1.35 Mbit/sec	10.1 MB	100.1 MB
16	48 kHz	1.46 Mbit/sec	11 MB	110 MB
24	96 kHz	4.39 Mbit/sec	33 MB	330 MB

A standard audio CD is 16-bits with a 44.1 kHz sample rate. A blank 700 MB CD would hold about 80 min of stereo audio. As you can see from the chart, a 24 bit/96 kHz recording takes up at least three times the amount of storage space compared to a 16-bit/44.1 kHz recording.

Lossy vs. Lossless

Many beginner audio engineers are confused about whether or not they should make beats or use samples with lossy files (MP3s) rather than lossless files (WAVs) that are a higher quality audio file format. A professional uses lossless files over lower quality audio files, lossy files, almost every time.

The two main differences between lossy and lossless audio files are storage space and audio quality.

A lossy file is created for compressing sound, image, and video. MP3s and Ogg Vorbis are lossy formats used in audio. Lossy compression is a data encoding system that compresses data by throwing out information deemed "unnecessary" to decrease file size. When it comes to sight and hearing, our minds easily fill in any of these missing gaps of information so we can see an image or hear a sound regardless of any small errors or inconsistencies. Lossy formats take advantage of a human's ability to "fill in the gaps." For instance, a low-resolution digital photo may be pixilated, but the image is discernible. We only need a limited amount of information to be able to process an image or sound.

File compression is meant to be transparent; however, if a file such as an MP3 is compressed below a certain limit, these gaps will be noticeable. A high-quality MP3 bit rate would be 256–320 kbps and a low-quality MP3 would be in the 100 kbps range. One of the biggest arguments against using lossy files is that once you compress the file down, you are unable to recapture the original information that was discarded. The advantage of using a lossy over a lossless format is that a lossy format provides a much smaller audio file. Lossy files are helpful when sending a large number of audio files by e-mail, or uploading audio or video to the internet. Lossy files are not intended for storing final mixes or high-quality audio files.

A lossless file is also compressed, though unlike a lossy file, when it is re-constructed, the lossless file is completely restored. No information is deleted and it is considered a perfect copy of the original. A lossless file will maintain the integrity of the original audio signal and does not leave it up to you to fill in those gaps. As demonstrated in the chart seen earlier, lossless formats offer the highest quality audio but take up quite a bit more storage space. AIFF (.aiff), WAV (.wav), and FLAC (free lossless audio codec) files are examples of lossless formats. Lossless files are used in professional recording environments.

In short, think of quality (lossless) vs. quantity (lossy). Lossless formats are more suitable for audio enthusiasts and professionals specifically interested in maintaining the integrity of the sound quality. The average consumer with a cell phone, computer, or iPod requires the convenience and storage space provided by a lossy format.

COMPUTERS AND AUDIO
Making Music With Computers

In recent years, computers have become an integral part of music composition, performance, and production. Although computers have been used to make music since the 1950s, recent advancements in technology have allowed engineers, musicians, and producers virtually unlimited control and flexibility when it comes to creating, editing, manipulating, and recording sound.

Musical Instrument Digital Interface, or MIDI, was developed in the early '80s as a data protocol that allows electronic musical instruments, such as a digital synthesizer and drum machine, to control and interact with one another. As computers became increasingly popular, MIDI was integrated to allow computers and electronic musical devices the ability to communicate. Over the past decade, the term MIDI has become synonymous with computer music. Now most commercial audio software has MIDI integration.

In addition to device communication, MIDI can be used internally within computer software to control virtual computer instruments such as software sequencers, drum machines, and synthesizers. Songwriters and composers who use virtual scoring software to create sheet music also use MIDI.

In many ways, MIDI works much like a music box or the player pianos from the early 20th century.

In this example, the player piano uses a built-in piano roll with notches that trigger the appropriate notes on the piano as the roll is turning. MIDI works in a similar fashion. Let's take a look at a simple MIDI keyboard controller.

This controller does not make any sound on its own; rather, it connects to the computer and signals to the software which key is being pressed or which knob is being turned. Each key, knob, and button is assigned a unique MIDI number that allows the computer to differentiate which note is being played (pitch), how long the notes are held (envelope), and how hard the key is pressed (velocity).

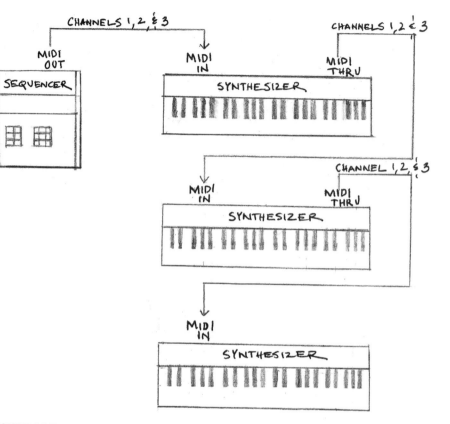

FIGURE 11.5

FIGURE 11.6

FIGURE 11.7

This information gathered from each key is stored in a MIDI note packet that is sent to the computer whenever a key is pressed. When the computer software receives the note packet, it unpacks and translates the information, assigning the incoming values to a virtual software instrument such as a synthesizer plug-in. The software then plays the appropriate pitch for the correct amount of time, and with the correct velocity. Additional controls on the MIDI keyboard such as knobs, buttons, and faders can be used to control virtual controls on the software instrument.

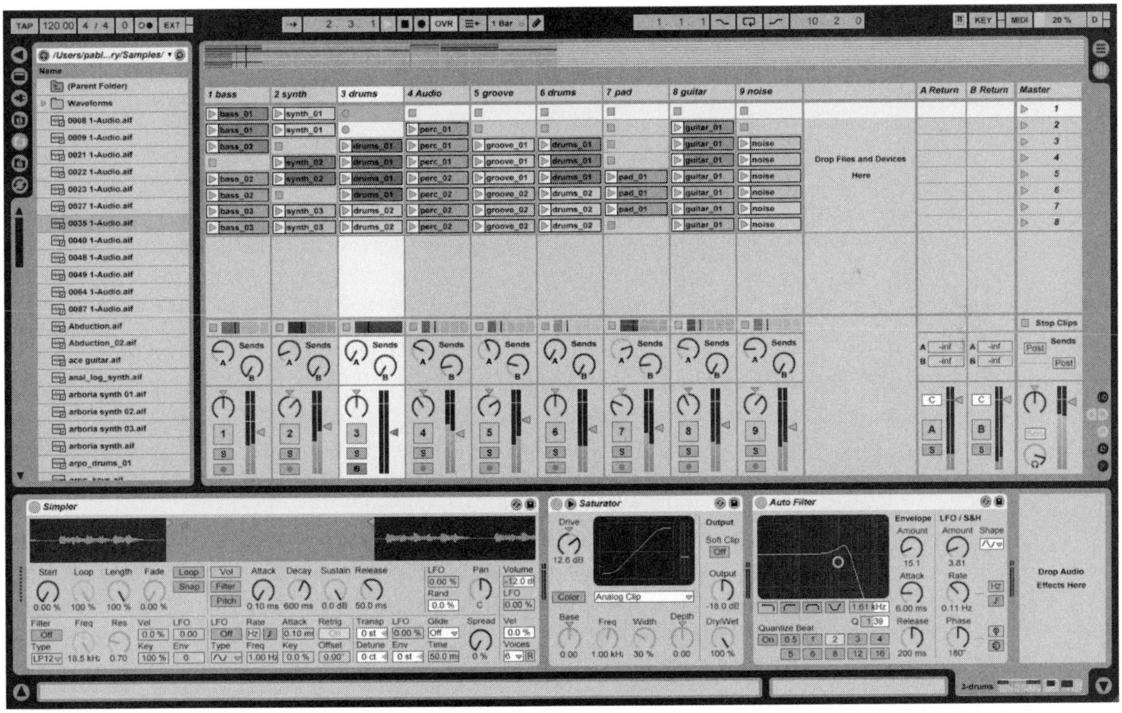

FIGURE 11.8

In addition to allowing real-time performance, certain music software programs have exclusive MIDI tracks that can record the MIDI information as a sequence, allowing of additional editing and arrangement of notes and controller information.

Since the early '80s, MIDI has been the industry-standard protocol for communication between electronic musical devices and computers. However, it has more recently been used to allow computers to control hardware devices in the real world such as lighting grids, video projections, and even musical robotics. Additionally, newer data protocols such as Open Sound Control (OSC) allow of faster and more flexible computer/hardware integration utilizing the latest in both Internet and wireless technology.

Software Choices

You want to start recording on your computer. With so many options of software, A/D converters and such, which set-up is best for you?

Here is a list of some of the most commonly used recording software/programs and associated equipment:

DAW (unless notes, software OS compatible with both PCs and Macs.)
GarageBand (Mac only)
Nuendo
Ableton Live
Pro Tools
Cakewalk Sonar
Adobe Audition
Logic Pro
Cubase
Digital Performer (Mac only)
Analog-to-Digital Converters
Apogee Rosetta
Lynx Aurora
Avid
Presonus
SSL
Scoring Software
Finale
Sibelius
Sequencers
Reason
FL Studio (PC only)
Acid Pro (PC only)
VST Plugins (Virtual Studio Technology)
Native Instruments
Sonalksis
Waves
Advanced Synthesis

NI Reaktor
Max/MSP
C Sound
Supercollider
ChucK
OSC
TouchOSC iPhone App
MIDI Controllers
Korg
M-Audio
AKAI
Roland
Novation
Monome
Livid Instruments
Lemur

Pro Tools is one of the most popular software programs used in professional recording studios. As with anything, people have their opinions, both good and bad, for this product. Whether you like it or not, knowing a little bit about Pro Tools isn't going to hurt, especially if you want to be employed in music production. That's not to say you shouldn't explore the other software options, in fact, I urge you to find what works best for your project and sounds best to you. Many engineers swear by Nuendo, Logic, and other digital audio software. I have found that once you understand signal flow and other basic audio technology, it is fairly easy to jump from one digital audio software program to another.

The Best of Both Worlds

As you will see in Chapter 14, FAQs answered by Pros, many professionals use a combination of analog and digital equipment.

Many professionals know that it isn't a matter of analog versus digital, but picking and choosing the best gear from both worlds. Many studios have Pro Tools, Nuendo, Logic, or other software technology along with an analog option like a 2-track or 24-track tape machine to color or warm up a signal. Studios with digital set-ups often use an analog front end. This front end can include tube preamps, compressors, and analog EQs.

There will always be people that insist that one recording medium is better than the other. However, many engineers stand somewhere in the middle, combining the best of both worlds and using what works and sounds the best to them, or is appropriate for their particular project.

Don't forget to check out Chapter 14's FAQs answered by Pros. Here you can find out what a group of select recording engineers think about the future, including their take on Pro Tools, analog, and digital audio.

CHAPTER 12

Now That I Am Dangerous, Should I Get An Internship?

As a Professor, probably the most common question students ask is about internships. It is understandable to want to get into a recording studio as soon as possible. However, there are real steps to take before you get there. Learning the craft is your first step. An internship is a way to get your foot in the door. It is not to teach you audio basics. This is an important distinction.

When I did my internship I had already taken several audio and media-related courses. Not to mention the live and recording sound experience I already had. I was a junior at the University of Texas majoring in Radio–Television–Film, and I had been a professional musician for about seven years. By the time I applied for an internship, I knew my way around a mixing board and I certainly wasn't a beginner. The more fundamental knowledge and experience you have the more likely your internship will be successful.

To illustrate my point about understanding the basics in the field before pursuing an internship, this is one of my experiences. After interning for a few months, the owner, who also happened to be the head engineer, said he needed to run to the bank and asked if I would do a few overdubs for the session he was working on. I successfully did the overdubs, didn't erase anything, and when the owner returned, he was pleased. I gained his trust. He knew he could leave his studio in my hands. Next thing I knew he was handing me sessions so he could run errands, play a round of golf, and eventually, take a much needed vacation. What if I had failed at recording the overdubs and erased tracks or wasn't capable of taking over the session? Maybe I wouldn't have been trusted enough to be asked to even help on future sessions. The lesson is you need some basic skills before starting an internship.

THE VALUE OF A GOOD INTERNSHIP

What an Internship Did for Me…

Everything!

There are many ways to gain knowledge about recording. To me, there is nothing more valuable than a good internship. This is especially true in the audio industry where experience is one of the most important requirements… An internship will give you much needed insight and experience and expose you to the real world of making records. Like I said, I had a few basic audio skills before I started my internship. However, I had no idea about so many other areas of engineering! In fact, I had no idea that I had no idea!

Things I learned in the four months of my internship:

- How a session flows from start to finish.
- The big picture and overview of the entire studio experience.
- The business side of a studio.
- Fee structure.
- How to communicate with musicians, musicians' friends, musicians' wives/husbands/significant others, producers, other engineers, the pizza delivery driver…
- How to listen more closely to the person actually paying for the recording, instead of listening to others that aren't as closely involved and not paying to record.
- Why classic gear is so classic?
- The look and sound of an U47, and why I can't wait to own one!
- The different sounds of classic compressors: LA2A vs. 1176 or a Tube Tech vs. a dbx165a.
- How to roll a mic cable properly? A very important skill! Make sure you check out the mic rolling video that accompanies this chapter.
- Technical skills: subtractive equalization, stereo miking, mic placement, and mic uses.
- Different genres of music.
- Making contacts with people from the local, regional, and national music scenes. It isn't who you know but who knows you!
- How to listen.
- When to talk.

- Perhaps more importantly, when not to talk.
- Basic signal flow.
- And finally, how much I didn't know!

I needed a backup plan to my drumming career, and I enjoyed running recording sessions and working with sound. My internship gave me connections to one of the best studios in Austin. It launched my professional career in music production eventually leading to me to teaching audio and writing this book.

My internship also had a direct effect on my professional drumming career. My band had only been together for three weeks when we made a demo that the right people heard. Being a handy intern gave me the perk of free studio time, which I used to record my band. Access to a professional studio allowed us to record the demo for free, which led to us being signed to a major label. Long story short: we cut a demo, got signed, and I spent the next 8 years touring around in a van with a bunch of dudes. It is hard to overestimate the value of a good internship.

Q & A
What Should I Expect to Do as an Audio Intern?

Keep in mind that this is a very competitive business. The owner or person who runs the studio wants you to prove that you deserve to be there more than all the other people that would love to be hanging around their studio. With that in mind, expect to do anything the studio owner would do: clean the toilet, wash the dishes, vacuum, clean up after a session, get burgers, gardening, go to the music store, you get the point – it's not always glamorous. Doing these things just proves how much you want to be there and shows your appreciation and dedication. If you think you are above these things then an internship is not for you. Don't be a whiner! Or a Complainer! Or a downer! (Now if all you are doing is cleaning toilets and you're not getting much else out of it then maybe this isn't such a great place for you to intern.)

A good studio owner will reward you with some free studio time or at least get you involved in sessions by having you do things like move mics, roll cables, perform patches, burn CDs, and other assisting duties. The chances are the more you contribute around the studio, the more likely they are to want and need you.

Keep in mind that the studio business thrives by word of mouth. Studios often expect you to help promote the studio and find potential clients. This can involve going to local music shows, handing out business cards, and getting out the word about the awesome studio where you are interning. It helps to be slightly extraverted. But you certainly don't want to sound like a used car sales person (no offense used car sales persons). Don't be too pushy or overly aggressive when promoting the studio.

Many studios use interns to help build out new rooms and provide labor on other construction projects around the studio. This can be a great way to learn about studio construction, acoustics, and sound isolation. My students always learn a lot from these experiences.

What Does the Studio Get Out of Me Interning There?

Free labor, plus a new person who can help promote their studio. See above for more details.

When Should I Ask Questions About Things I Don't Understand?

Never interrupt a session in progress. The musicians couldn't care less about your audio education. They are paying good money to record, not to have the engineer teach you audio on their dime. At the best, you are a minor nuisance to the band that can run and get them beer and at worst they ask you to leave because you are ruining their awesome vibe! The best time to ask questions are generally at the end of a session when it is just you and the engineer. Select one or two things to pick his or her brain. Make it brief!

How Many Hours Should I Expect to Intern a Week?

Most internships are for colleges or university credit, and they usually have their own specific requirements. Where I teach, our internships require 10–15 hours per week for a 16-week semester, somewhere between 144 and 288 contact hours per semester. Many studios may require you to work many more hours than your requirement. During my internship, I would spend up to 60 hours per week at the studio, even though I was only required to work 14! I loved it, couldn't get enough of it – I knew this is what I wanted to do, so I enjoyed the extra time. I remember this cushiony old brown chair in the corner of the control room. After I would finish daily studio duties, I would sit in that old chair for hours, absorbing everything. I made sure not to disrupt or talk when I was hanging out so I wouldn't annoy the band. Putting in those extra hours quietly observing prepared me to take over that first session. This is where I learned how things were communicated with musicians, the general workflow of a recording session, why a specific piece of outboard gear was picked over another, and much more. This was invaluable. You can only get this experience through an internship.

Why Should I Work for Free?

An internship is a life experience you cannot get from a book, a class, or any other resource. Audio engineering is one of the few careers that most people still apprentice. You will learn how a professional session runs from setup through mixdown. The way a session flows from start to finish is very important. The more efficient you are as an engineer, the more money you will save your client. Remember, time equals money! And most musicians aren't rich!

How Do I Get an Internship?

I got an internship because I happened to be the drummer on a session at the studio that would eventually sponsor my internship. I asked the engineer/owner if he needed an intern. He said he would like an intern. I started shortly thereafter and signed up for the class at my university to get actual college credit for it. So, being a musician got me in the door and going to college for audio gave me a reason to ask for an internship. At the college where I teach, we offer an internship class for graduating students. We provide a list of some local studios that take interns. Often students find their own studio the same way I did. They record at the studio, like their experience and asked the studio if they needed an intern. Research the studio that best suits you and do whatever it takes to get in there.

Can I Give My Opinion?

No! No! and No!

A few of my studio owner friends have told me nightmare stories about interns who would say things completely out of line to their clients. These studios no longer take interns. Don't be the fool that ruins internships for others! As an intern, you are there to observe and help whenever you are needed. You can talk to clients but don't blab their ears off and give them your opinions! You are not there to tell the band that was a good take or to tell the engineer how you would do it differently. Just because you have a certificate from an audio school and you own an MBox doesn't mean you are an expert. Remember, it is a privilege to be there. The only exception is if the musician directly asks for your opinion.

Do I Need a Resume?

Having a resume certainly doesn't hurt. However, you may not have much to put on a resume at this point. Having samples of things you have recorded on a CD or thumb drive may be as valuable as a written resume. If a written resume is required, be professional and emphasize what you have worked on. Keep in mind that actual recording experience weighs much heavier than education, book knowledge, and a recording school certificate.

INTERNSHIP'S DOS & DON'TS
Do

- Be on time or early. Your job competition is/are often musicians. They are inherently flakey and irresponsible. Showing up on time can make you stand out. I credit this detail for much of my success.
- Become invaluable to the studio, so they can't live without you. This will make the studio want to keep you around well after your internship.
- Makes friends at the local music store, professional tape supply company, and other related studio businesses. Since you will likely be a "runner," you will meet people associated with the business while picking up necessary supplies. I have many life-long business contacts and friendships from these early experiences.
- Help unload gear when the band arrives. This is a quick way to make a few friends.
- Ask if anyone needs anything. Offer to make a food run.
- Answer the phone when needed.
- Be friendly and introduce yourself.
- Complete any tasks that are expected of you.
- Enjoy what you are doing. If this gig is for you, menial tasks or other non-engineering task will keep you humble and hungry to work when you get the chance.

FIGURE 12.1

- Take initiative.
- Thank the owner or engineer for letting you help, learn, and observe.
- Pay attention to the interaction between the engineer, musicians, producer, and other folks involved in the making of music.
- Be open minded and prepared to absorb everything going on like a sponge.
- Be professional.
- Practice personal hygiene. No one wants to sit in a control room with a stinky intern!

Don't

- Stand around. Instead, make yourself useful. Again, you want to be invaluable.
- Act cocky, rude, or give your opinion on how you would do things.
- Gossip about clients.
- Text during a recording session. In fact, don't text at all unless it has something directly to do with the session, like placing a pizza order.

- Take pictures or ask for autographs.
- Ask what you should do next. Be proactive.
- Be overly friendly and talk people's ears off.
- Be late ever!
- Take it personally when a client, musician, or engineer says something rude or snaps at you.
- Ask for free studio time at the beginning. This privilege usually comes in a matter of time when you have gained the studio operators trust.
- Ever flake on any obligation, especially if the studio is counting on you to complete a particular task.

▲ TIP

Finally, don't be a know it all! Be humble.

EXAMPLES OF INTERNSHIP ADS

An internship can be found through a college, university, or audio school program while you are in attendance. Your local paper, online sites, or trade magazines also have ads for internships. These three advertisement/applications are examples of what a studio might ask and expect of an intern.

First Ad

STUDIOS RECORDING STUDIO INTERN

Recording Studio Intern Application

This is a 3-month long, non-paying internship position, consisting of various "runner" jobs. Completing an internship at _____ Recording and Mastering can help you understand the bigger picture of the production and studio business. It will also help narrow your focus on which aspects of pro audio production fit you best.

Name:

Address:

Phone:

E-mail:

Which recording school have you attended/are you attending?

Have you interned before?

Are you a musician/in a band?

Do you own and/or use Pro Tools?

Is your focus on creating music or working in post production?

(Continued)

Home many days per week can you commit to?

How did you hear about _____ recording and mastering?

When can you start?

Salary: Lunch and 10 dollar gas

Second Ad

Very low-key studio in the _____ area is looking for interns that meet requirements below. This is a professional studio where platinum records are recorded daily; half of the hit songs playing on the radio right now have been worked on in this studio. So no clowns looking for autographs please.

PLEASE READ ENTIRE POST BEFORE REPLYING!

1. Must know how to work in a professional studio. Not an MBox setup, a full HD rig with ssl consoles, patch bay, drum mics, and $15,000 mic.
2. This is a non-paid internship. Do not come in here with expectations to get paid directly from the studio ever! You're here for experience and to get connected in the industry, and if you get in good with the engineer you might even get an assisting credit or two.
3. You are here to work, not to be on your cell phone, laptop, or talk about your beats or songs all day. If you have a problem with being a hard worker do not bother sending a reply.
4. You will never approach any one with demo or beat CD, unless specifically asked by the client. Unless someone asks what you do, they do not need to know.
5. Must have a full availability for at least 5 days a week. We are only looking for a few people to fill the spots. Expect long hours when needed. Sessions sometimes run non-stop 2 days straight.
6. Must live near by. Half an hour away the most. We need someone to rely on who's not gonna take 2 hours to commute.
7. Must have multiple internships at other studios. (If all you ever did was runs and label CDs don't bother applying.)
8. Expect to do everything from cleaning bathrooms, answering phones, to tracking platinum artist and producers and if clients ask you to do a back flip – do it.
9. The word "no" must not be in your vocabulary. You should want to do everything that is asked of you.
10. No haters, thieves, or crybabys. We need people who can work together.
11. No getting star struck or asking for pictures or autographs. These are means for immediate termination.
12. You will not take anything from the studio. This includes cables, headphones, samples left on an mpc, demos lying around, and especially any sessions on the house drives.

So to sum it up: we need one or two interns that are prepared to walk in here and assist on sessions for high-profile artist. If you think you meet the requirements, reply with a full resume and a paragraph telling us why we should consider you. If considered, we will schedule a phone interview followed by a trial session at the studio.

Third Ad

Industry: Entertainment

Location: South _____

Position: Audio/Studio Engineering Internship

Salary: This is an internship, learning experience with potential to make money

About Us: A brand new recording facility in South _____. We are in the position to be selective about the projects we take and will only put our attention and resources on sessions we believe in. We work in a low-key environment and treat everyone like the "boss," including the intern.

Duties Include: Knowing the studio inside and out, understanding signal flow, working a patch bay, setting up microphones for drums, amps, etc., to enable you to eventually engineer or assistant engineer sessions, keep studio equipment maintained, cater to reasonable client needs, and have fun in every session.

Opportunities Include: Working with seasoned professional artists with hit songs, as well as up and coming talent. Helping to find great music to record.

What We Are Looking For: Talented, hard-working, dedicated, loyal, focused, creative, highly motivated, and reliable individual full of potential that will truly make the most of this experience. Come in with a professional attitude and a strong desire to learn. This is a very demanding yet highly rewarding, full or part time, unpaid internship, so please only those serious about starting or furthering a career in the music engineering field should apply.

Requirements: Graduate or student from an audio engineering degree program (8 month minimum, associate or bachelor degree preferred), solid foundation of engineering, knowledge of signal flow, Logic Pro experience, musical knowledge a plus, technical aptitude, good people skills, problem-solver, career focused, and a reliable car is a must.

Schedule: Flexible

Length: Flexible

To Apply: E-mail your resume to the following address; include a cover letter explaining why you want this position.

GET IT?

As you can see from these ads, much of what I mentioned in this chapter is directly in the ads. The more knowledge and experience you gather before you attempt an internship, the more likely your internship will be successful. I can't stress how much an internship did for me. It helped me develop the craft, provided me with a mentor, and gave me insight into the crazy, yet entertaining, music business. Unlike other methods to learn audio engineering, an internship provides you with "real" experience. You can't get this unique experience from a book or the Internet. If you are lucky and do good work, you may even get a job out of the internship.

8 How to Roll a Mic Cable

▲ TIP

Rolling a mic cable properly will give you instant credibility in the studio or in a live sound situation. This is a must-have skill before starting an internship. Rolling the cable properly also protects the cable and will make it last longer. Not rolling it properly will tip off other engineers that you have little or no experience. This video demonstrates the basics of rolling a cable properly.

CHAPTER 13

Jobs. What Can I Do With These Skills?

As mentioned in Chapter 4, if you decide to become a recording engineer you are likely to be self-employed. Very few places now offer staff positions in music production. What this means for you is there is not a guaranteed paycheck every month. Your fee will typically be separate from the studio or your employer. The exceptions are with some churches, music venues, and live sound production companies. Working from job-to-job is one of the many reasons you will often take on more than you can handle because you won't know when your next gig will happen. Most freelance arrangements do not include any type of contract. Typically, these arrangements are verbal agreements between the engineer and the management or owner of the venue or studio. With most studio arrangements, the engineer is paid by the band, whereas venues usually pay the engineer. Many engineers have other skills they can use when the projects aren't rolling in. These skills may include building websites, booking or promoting bands, repairing equipment, carpentry, electrical work, and even teaching audio-related classes, like me.

In the late 80s, when I began my career, there were very few home studios. The only affordable options were to record on a cassette 4-track, which was only good enough for demo quality, or go to a commercial recording studio. In the early 90s, the digital revolution brought an affordable and less cumbersome recording setup for the hobbyist and home studio. This setup involved the 16-bit ADAT that was often paired with a Mackie console. Alanis Morrisette's hit record, *Jagged Little Pill*, was recorded in a home studio with this setup. Since then, the home studio has continued to grow and you can now produce professional quality recordings without going to a large commercial studio more than ever.

Fast forward 20 years and home studios have continued to replace the traditional, larger commercial studios with the digital audio workstation (DAW) taking over where the ADAT left off. Another reason home studios have steadily increased is because recording gear has continued to become more affordable and specifically designed for home use. It is common to track drums at a larger studio and finish the rest of the tracks back at a home studio. In fact, many times, all the recording and mixing are done in the home studio. Many engineers choose to have home studios rather than work out of commercial studios. Many engineers are flexible, owning, and running a home studio, in addition to engineering and recording projects at outside commercial studios. This flexibility provides a freelance engineer with more opportunities to work more places and meet more clients.

LIVE SOUND ENGINEER

Having toured around the country, I can safely say a lot of local live sound people aren't known for their punctuality or great communication skills. I have also run across sound people with bad attitudes. They acted as if they were doing me a favor by miking my drums, instead of treating it like part of their job. This is not to say there aren't great local sound engineers who behave professionally, because I have met plenty. If you can excel in these areas, where many do not, and also have a decent ear, you will likely have plenty of live sound work, at least in my town. Many of my friends and students entered music production by first running live sound. Unlike a recording engineer, there are generally more opportunities in live sound. This is because more people perform music on a regular basis than need it recorded. I ran sound off and on for many years, often with the offer of more gigs that I could handle at one time. This may have been because I was always on time, easy to work with, and I even knew something about sound. You too will be highly sought after if you are well liked, dependable, and knowledgeable about live sound. If your ultimate goal is to record music, live sound is a great way to meet potential clients to record. As an independent engineer, it will be up to you to find people to record, and running live sound will provide you with an endless list of potential clients.

Venues that have live sound include:

- Churches
- Special events
- Live music venues
- Festivals – In Austin alone there is the infamous SXSW, Austin City Limits, Fun Fun Fun Fest, Pecan Street Festival, and many more. There are festivals all across the country, not to mention all the international festivals.
- Touring bands

- Theaters
- Comedy clubs

Benefits of running live sound:

- More available job openings
- Easier to get started
- Allows you to work on your engineering skills
- Great way to meet more musicians and industry people
- Provides immediate performance feedback

What Should I Expect to Do if I Run Live Sound?

Depending on the size of the venue and whether you work for the venue changes what your duties may involve. If you are hired by the venue, you will be considered the house sound person or in-house engineer. You will be responsible for setting up and tearing down the sound equipment. At a smaller venue, you may even pull double duty as the sound engineer and the door person or bartender. Whether you are the in-house person or not, your main duty is to set up and run the sound for the band. This usually includes miking the musicians, setting proper mic levels, preparing monitor mixes, and making sure the show starts and ends on time. If you are the in-house engineer, you are typically part of the club staff and additional duties, not related to running sound, may be expected. A freelance engineer is typically hired by the band to be their personal sound engineer. A freelance engineer typically shows up for their band's sound check to preset levels and become familiar with the venue. Unlike the in-house engineer, they will only be responsible for mixing the band that hired them.

How Do I Get a Live Sound Gig?

It is common in music production to shadow or assist an experienced engineer. Assisting an experienced audio engineer is a good way to learn more and get your foot in the door. If you don't know a sound person that you can hit up to assist or cover for, find a venue you like. Scout it out. Make friends with the sound person. Find out if he or she ever needs someone to cover their shift or if they know of any other sound gigs that are looking for help. Offer to help them set up and tear down for free. This will give you an opportunity to become familiar with the venue and you are building a relationship with the staff. Many live sound engineers are also musicians and may need their shift covered when they are playing or touring.

What Should I Expect to Get Paid?

Pay varies, depending on the type of venue where the gig is taking place.

There is a wide rage in the pay scale for audio engineering. As we've covered, you may work for free at first, to gain experience. Most live sound gigs start around $10–$15 an hour, with a shift ranging from 1 to 16 hours long. A typical weekend night gig in a small Austin club would pay about $50. If the venue is a restaurant, food may be included with your pay. Weekends generally offer better pay because the shows are likely better attended and there is more money to go around. Most gigs start around 9 P.M. and end around 2:30 A.M. or within an hour or so after the club or venue closes. As you become more experienced, you will probably move onto working for bigger and better venues, which are likely to pay more, have better gear, and be more professional. If you do get a job at a larger venue, you are likely to start off on monitors and not be in charge of front of house (FOH) sound. If you love music you will hear and see plenty of it, which is a perk.

Churches generally pay a little better, and you hopefully won't leave work smelling like beer and cigarettes. A church gig may start as early as 6 A.M. and go to about 1 P.M. It can be an hourly paid job or per shift with the pay varying from $10 an hour up to $75 an hour. The pay depends on the size of the church, the church's budget, and the importance of sound with the church service. The downside for some is the gig usually starts early, so if you aren't a morning person a church gig may not be for you.

Local music festivals and other musical events not taking place in an established venue are staffed through a local sound company. Local sound companies tend to provide a portion, if not all, of the sound for festivals and other community events. You won't have to worry about hustling work if a sound company employs you. They pay either an hourly wage or per project. A huge benefit for working for a sound company is that you won't have to worry about finding clients or where your next paycheck will come from.

Going on tour with a band or an artist usually pays the best money, as far as live sound gigs are concerned. There is a trade off: it will be a 24-hour-a-day job and you may not get much time off. The smaller the band, the more likely you will do more than just run sound. You may also help with selling merchandize, driving, some tour managing, running monitors, collecting money at the door, or whatever else they may need. With larger touring acts, the FOH engineer will only engineer and will not have to do the other miscellaneous jobs that come with touring with a smaller act. As it is in other venues, you will start on the monitors and work your way up to mixing the front of house.

If an artist adores you, you may be put on a retainer. The retainer fee guarantees you will be available to this particular artist(s) first. The retainer fee also guarantees you some cash when the band or artist isn't touring. The fee is usually a portion of your typical touring pay.

▲ TIP

To succeed at live sound:

- Show up on time.
- Be prepared.
- Don't leave the mixing board unattended to drink beer and play pool. You never know when mic *feedback* may occur. Stick around and be a pro.
- Help the band setup and tear down, especially if the event is running behind.

- Take pride in your work and don't be lazy. If a monitor needs moving so that someone can hear better, move the monitor. Be proactive!
- Don't ever panic if things aren't going your way or you feel rushed to set up the sound.
- Have a good attitude.
- Did I say show up on time?

RECORDING STUDIO ENGINEER

Unless you have mixed more than a few bands, a job at a recording studio usually starts as an internship. As discussed in the previous chapter, don't attempt an internship until you have knowledge of audio basics and feel you could, with a little more time, run a professional recording session.

Almost all studio engineers are self-employed, as very few studios actually have staff engineers on their payrolls anymore. Engineers are typically paid separately from the recording studio. For bands recording in a studio, the studio will generally provide rates with and without an engineer. Let's say a studio's rate is $500 a day with an engineer or $250 a day without an engineer. The band hires you as an independent engineer. You and the band agree to book the studio for 3 days ($750 studio time). You would charge the band whatever fee you feel comfortable charging on top of that $750. If you were to charge $150 a day you would end up pocketing $450. In this case, the band would pay $750 to the studio and then pay you $450 for engineering for a total recording cost of $1200. The band saved $300 by hiring you as an outside engineer versus using the engineer provided by the studio. Typically, bands will settle up with the engineer after they settle up with the studio. This all depends on your agreement with the studio, but you always want to make sure the band covers the studio. Eventually, the more time you book with the studio, the more likely they are to make you better deals in the future. You will be able to increase your fee as you gain more experience and develop a working relationship with the studio.

▲ TIP

Studios generally have a little wiggle room with their quoted rates especially when money is put on the table!

Figure 13.1

I have found that the more I like a band and want to record them, the more likely I am to work below my typical rate. If a band is good and isn't signed or represented by a huge management company, they are possibly short on funds and can't afford to spend much on studio time. You might want to consider working with a band like this for experience and to add to your resume and to get your foot in the door with a potentially successful band. I have made good deals with bands with a verbal understanding that when they do get a real budget to contact me to work with them. Working on a demo or an EP for less money will pay off, when you get the call for the bigger budget gig in the future.

You will likely be self-employed, so I suggest you ask the band or artist the following questions I previously mentioned in Chapter 9 before you take on a project.

Four things to know before you record a band:

1. What is the budget?
2. What type of instrumentation will the band be using?
3. What is the purpose and length of recording?
4. What are the band's expectations?

Even though you will probably take almost any session, you can get when you are starting out, asking these four questions can give you a better idea of the project and alert you to any potential red flags.

The budget will give you an idea of where the band can afford to record and how much time to spend on each stage of the recording. It will also help you determine if the band has realistic expectations (number four on the list).

Instrumentation helps you also determine if the budget matches the band's aspirations. A session recording a singer/songwriter with an acoustic guitar and a vocalist is much different than a session recording a full band. Also, knowing the instrumentation will allow you to mentally prepare for the recording session.

What is the purpose of the recording? A demo to get shows? A self-released CD? A commercial recording on an indie label? What is the length of the recording? Two songs? Ten songs? Answers to these questions can help you determine if the band's budget aligns with their expectations. Make sure to get the approximate time of each song. I took a session once with a band that wanted to record five songs. I had no idea their songs were 12–18 minutes each! Obviously this band's jazz jams were very different from the 21 songs I recorded for a punk band whose complete album was less than 30 minutes in length!

Are the band's expectations realistic? Can you accomplish the recording that the artist wants for the money and talent of the artist? If not, you will want to either resolve this or turn down the project. Be honest with the artist or band if you don't believe their expectations are achievable and intelligently lay out an argument why. Don't get defensive and stick to the facts.

One of the coolest things about being an independent audio engineer is that you can work at almost any studio you desire. You only need a band or artist with a budget.

ASSISTANT ENGINEER

An assistant engineer assists the engineer or producer with the session. An assistant engineer is more commonly found in larger budget sessions. Often, the assistant engineer is the main engineer or an advanced intern at the studio that picks up the gig for some extra money. Generally, the assistant engineer is very familiar with the studio, its gear, and the recording spaces. This is helpful to the outside producer or engineer who typically requests an assistant because they are unlikely familiar with the studio. The assistant engineer, also known as the second engineer, will help with patching, moving mics, computer operation, or any other duties to help the session run smoothly. Assistant engineers usually get paid a flat fee for their services. Personally, I have been paid anywhere from $50 to $200 a day to assist on a session.

MASTERING ENGINEER

As described in Chapter 9, mastering is the final stage of most recordings for an engineer. Mastering engineers either get paid per project or per hour by the band or the record label. The pay scale varies depending on your experience and reputation. If you are very detail oriented and would be considered an "audiophile" by your friends, being a mastering engineer may suit your personality type. My mastering engineer friends like the fact that they generally spend a day or two, not a month or two, per project. This is because mastering doesn't require all the production details of the performance, recording, and mixing of the album.

Most mastering engineers have a neutral sounding room and many sets of high-end speakers to accurately monitor sound. Mastering engineers also have a favorite analog compressor and/or EQ to help them perform their duties at the highest level. As a mastering engineer, you could work on and complete hundreds of recordings a year. This isn't really a position where most engineers start but a position experienced engineers move in to. Here is one suggested reading on mastering: *Mastering Audio, the art and the science,* 2nd Edition, Bob Katz, Focal Press 2002.

POST-PRODUCTION ENGINEER

Although this book mainly deals with music production, post production often includes music and engineering skills. Post-production work deals with audio as related to film and video. Post work could include recording voiceovers, dialogue, and mixing audio for picture. Post work could also include foley work, like recording footsteps, moves, and other specific sounds such as bones being crushed or the sound of gunshots. More recently, post-production work includes audio for phone apps and video game sound. These are two emerging areas of audio production. If you want more information, there are plenty of books and other resources that specifically deal with post work.

RADIO PRODUCTION ENGINEER

Radio production jobs can be in public, commercial, Internet, or cable radio. You may be expected to do some of the following:

- Record, edit, mix, and master program audio and provide quality control, ensuring that the highest quality content is delivered to audience.
- Operate DAWs, digital audio recording devices, complete CDR and DVD-R authoring, and digital distribution systems.
- Work collaboratively to develop the overall production and sound quality for new daily programs.
- Manage program and session archives.
- Coordinate, plan, and engineer field and remote productions.
- Oversee daily studio operation for local, state, and national programs.
- Creatively choose and apply the use of music, sound elements, and interviews to the programming.
- Record, mix, and edit programming for national distribution as assigned.

The pay scale in radio production varies. The good news is radio production jobs are one of the few production areas that you aren't likely to be self-employed, but instead have a salaried position. The hours will be more traditional and you may even get some type of benefits.

PRODUCER

Not all recording sessions involve a producer, although one may be needed. As an engineer, this means you may get stuck producing by default. Most producers, if they weren't audio engineers first, have some technical background or knowledge. Producers are usually paid per song or per project by the band or record label. The pay scale can vary from a "spec deal" to tens of thousands of dollars per song or project. Some producers may offer an artist(s) a spec deal. In this case, the producer is footing the bill for the project with a verbal or written understanding that if the song or album does well the producer will get a certain percentage in return. There is no standard deal or rate.

Here are the main duties of a producer:

- Song selection.
- Revising songs.
- Deciding on the type and the purpose of recording.
- Arranging the instrumentation and deciding on what instruments will be used.

- Assisting in or overseeing the preparation of lyric sheets or music charts.
- Selecting a recording studio.
- Selecting an engineer.
- Hiring and rehearsing musicians.
- Critiquing performances.
- Communicating between the artist and engineer.
- Guiding the budget process.
- Overseeing all phases of recording: pre-production, recording, overdubbing, mixing, mastering.
- Helping to shop the final product to labels, managers, etc…
- Overseeing the final projects artwork, design, liner notes, and credits.
- Finding a quality duplicator/replicator.
- Defining a market for the project.
- Assisting the band with publishing and mechanical licensing.

If a recording session doesn't have a producer, someone will end up performing these duties. Whether it is the audio engineer, the artist(s), or both, who ends up picking up the producing duties will depend on the project.

THE HOME STUDIO

With digital audio technology becoming more affordable, home studios have become a common place for music production. You can have a complete recording studio in your laptop. This was unimaginable not too long ago! Technology has made it possible to make recording at home better than ever. A home studio can make recording more affordable for the artist by eliminating the middleman – the recording studio.

For an audio engineer, there are many benefits of the home recording studio. You no longer have to pay a fee for studio time. This makes it easier to charge a little more for your services and still save the band a few bucks overall. I have a home recording studio that can handle most of the projects I am interested in working on. This allows me the ability to record or mix at an outside studio only when necessary, or when a change of pace is needed. I may mix at my friend's commercial studio to benefit from the automated Trident console and racks of classic gear or track drums at another studio for a larger, different room sound.

Another benefit of the home studio is not having to pay rent for space and decreasing your overhead. It is hard enough to make money when you are first starting out, and paying to rent a space on top of your normal living expenses only makes it tougher. This is a great way to kill two birds with one stone, or in this case, two rents or mortgages with one place.

Since recording gear has become affordable and easier to operate, many people believe that purchasing the latest software will allow them to produce quality recordings. Without a basic understanding of sound and recording, it is unlikely someone will record a quality, professional product. Professional recording engineers are often underappreciated for their ability to understand sound. That is, until a person with little or no experience attempts to make their first hit record on their own. Clients like this end up at a studio, seeking professional help, frustrated because they spent their recording budget on recording gear without considering you also have to know how to operate it and make it sound good. I'm not saying people can't get lucky, but good music production takes years of experience and knowledge to create.

If you start a home recording studio, make sure you have more skills than your average hobbyist. Just because you can work a software program or you have great computer skills doesn't mean you know how to record and mix a record. If having a home studio interests you, learn the basics, get some experience, and figure out how to make records *before* you consider yourself a professional. Purchasing the tools of the trade doesn't make someone automatically qualified to make quality recordings.

PROs and CONs of a home studio:

PROs	CONs
You never have to leave your house.	Strangers will be in your home.
You won't have to pay a studio for time so you can usually make more money.	You do have to equip and maintain a home studio, which can be expensive.
You can cut your overhead expenses by combining a rent or mortgage with your business costs.	Customizing your workplace can be expensive. Most larger commercial studios have better recording equipment and larger recording spaces than can be created for a home studio.
You can customize your workplace.	Neighbors may complain about noise or traffic or suspicious looking long-haired dudes walking around the neighborhood.

🜂 TIP

Tips on starting a home studio:

Feel out your neighbors and make sure there aren't city or neighborhood restrictions. You would hate to do a bunch of work only to get shut down or constantly harassed. Check with your city ordinances to find out if your area has specific restrictions.

Register your studio name or production company. This can be done at the city courthouse or online by registering your business as a do business as (DBA), sole proprietorship, or a limited liability corporation (LLC).

Open a bank account under your studio name.

Keep receipts of all your purchases. This will give you proof for potential tax write offs.

Start accumulating gear: mics, compressors, FX units, and whatever gear you will need, so you can spend less later and put the money you make toward your current living expenses.

Start collecting toys now: tambourines, shakers, specialty amps, a theremin or two, toy pianos, a real piano, and other unique, creative instruments. This can differentiate your studio from an artist who is debating whether to record elsewhere.

EQUIPMENT MAINTENANCE AND REPAIR

One person who is in constant demand is the person who can repair amps, consoles, tape machines, and any other studio gear or musical equipment. If you are good at or interested in basic electronics, this may be the career path for you. Fewer and fewer people are trained in basic electronics and circuitry, leaving more demand for people who can fix studio gear and musical equipment. Equipment maintenance is a job you can do out of your house or garage on your own terms. Of course, you would have to travel to recording studios and other businesses to do the troubleshooting and repairs. You could also look for work at music stores or specialty shops. In most towns, if you can fix a vintage amp, musical equipment, and studio gear, you are sure to have more work than you can handle. If you are dependable and highly skilled, you will be able to set your own rate and hours. To acquire these skills, you can take basic electronic classes or start tinkering around on your own. Some colleges offer degrees in Electronics. Learn about circuits, resistors, and capacitors. Purchase a soldering iron and practice repairing cables and other audio-related equipment.

You can purchase a kit to build a handmade synthesizer and learn a bit about circuits, circuit bending, and electronics at www.bleeplabs.com.

TAKE A BUSINESS OR ACCOUNTING CLASS

Part of the Commercial Music Management degree plan at Austin Community College requires students to take a Small Business and Accounting class. This makes sense, considering you will likely be self-employed and you will need these valuable skills to succeed.

Every semester my Audio 4 class visits local studios. One particular owner/engineer always advises my students to take classes in marketing and accounting. He is the first to admit that the business side of running his studio isn't his best asset, because he did not educate himself on how to do these things successfully. If you decide to run your own studio, you will want to be as prepared as possible for all aspects of running a business – not just the musical aspect. If handling and budgeting money isn't one of your strengths, you will want to address this issue. Consider what it takes to succeed as a small business: advertising, marketing, people management, money management – the list goes on. It is one thing to be a good audio engineer; it is quite another to be a successful small business owner. Colleges and universities offer classes in both these areas, business and accounting. Take advantage of them.

SHOW ME THE MONEY

It will be up to you to negotiate a fee for your services up front, as a freelance engineer. Don't wait until the project gets started to discuss your fee. You shouldn't talk money issues while recording and mixing.

If you are unsure how to set your own rate, ask other audio professionals in your area what they charge for their services. Adjust your fee appropriately: charge less than a person with more experience and charge more than a person with less experience. Every project is different, so expect a budget to also vary. You will have to be flexible with your fee if you want to work a lot. Don't sell yourself short, but you have to be realistic about what others in your area and with your amount of experience are getting paid.

Whether you get paid per hour, per song, per project, or per day, make sure you and the client are clear on your fee. Money and art don't mix, so never talk about money during the project. It is customary that most engineers and producers require 50% up front and the other 50% when the project is completed.

▲ TIP

If you are charging per project, be conscious that a client could drag a session on and on. Make sure the client is aware that your project fee doesn't mean they will have an unlimited amount of time. Clarify the project length and don't leave the time frame open.

No matter what area of audio engineering you pursue, the amount of experience you have will be one of the determining factors of what you get paid. Although a formal education is helpful, it isn't necessary. Most employers would rather hire someone with years of experience over someone with only an audio-related degree and no experience. If you decide to be an audio professional, don't do it simply for the money. Do it because you love music. If you stick with it, you will get paid what you deserve. Over time, if you show up on time, are motivated, positive, creative, and willing to do whatever is needed for the session or gig to go smoothly, you will be successful. As previously mentioned, this is a word of mouth business. If word is you are easy to work with, talented, and dependable, you will be on your way to a successful career in music production.

CHAPTER 14

FAQ's. Hear It from the Pros

A few years ago, I began compiling a list of the popular questions asked in my classroom and in the studio. I thought it would be valuable to ask a diverse group of respected audio engineers from around the country these questions, to gather their answers and opinions.

The questions in this chapter range from "What is the first microphone I should buy?" to "Which recording software do you use?" to "How did you get started?" The answers vary, giving us insight into what the Pros think and hopefully will give you confidence that there are many ways to achieve the same end result.

Below is the list of those engineers who participated. In addition to being audio engineers, many of the participants are producers, musicians, artists, professors, mastering engineers, technicians, and studio owners.

(AM) Andre Moran, Congress House, Austin, TX
(CJ) Chico Jones, Ohm Recording Facility, Austin, TX
(CS) Craig Schumacher, Wavelab, Tucson, AZ
(FR) Fred Remmert, Cedar Creek and Cherokee Recording Studio, Austin, TX
(GS) Greg Smelley, Marfa Recording Co., Marfa, TX
(HJ) Hillary Johnson, Senior Tape Op Contributor, Independent Engineer, New York, NY
(JH) John Harvey, Top Hat Recording, Austin, TX
(JW) Jim Wilson, Jim Wilson Mastering Boulder, CO.
(KM) Kurtis Machler, Million Dollar Sound, Austin, TX
(ME) Mitch Easter, Fidelitorium, Kernersville, NC
(MP) Mary Podio, Top Hat Recording, Austin, TX
(MR) Mark Rubel, Pogo Studio, Champaign, IL
(TD) Tim Dittmar, las olas recording, Georgetown, TX

GEAR

1. What Is the First Microphone I Should Buy?

Mitch Easter (ME): Get a dynamic mic that costs around $100, or a nice used one that originally cost that much new. You can record anything with it, and get pretty good results! Dynamics are tough and can be jammed against guitar amps, stuck in bass drums, etc., and they won't care. At the same time, they sound perfectly good on voices. Supposedly Bono and Tom Petty have done vocals on their hit records with Shure SM58s, which seems entirely plausible to me.

Chico Jones (CJ): Buy the mic that sounds best on your own voice or instrument. You may need a friend or bandmate to help you with this test. It's a great way to begin learning the subtle differences between mics – using the human voice. Do some research. Borrow a couple of mics first. Buy what you can afford. I think the Sennheiser MD 421 is a good mic that you won't outgrow too quickly. If that is too expensive then a Shure SM57 or Beyer Dynamic M88 will be handy as your collection grows.

Fred Remmert (FR): Well, that sort of depends on how much money you have and what you are trying to accomplish. If you're a millionaire and want to start a collection, then go find the rarest Telefunken U47 you can find. If you're a broke engineer starting out and want something you can use to record just about anything, get a Shure SM57.

Mary Podio (MP): The first mic you should buy is a mic that you think sounds good and that you can afford. Don't worry if it's not a U47, plenty of people have made great sounding records with inexpensive mics. Your greatest asset is your creativity. Trust your ears and make whatever you have work for you.

Andre Moran (AM): Shure SM57.

Greg Smelley (GS): The one you can afford right now. It really depends on what you will be recording. The decision between dynamic, condenser, or ribbon may be dictated by style of music. I do, however, think a Shure SM57 or 58 is worth having early on. Everyone is familiar with the sound of those mics, so getting a usable sound can be easier because you can get a familiar sound. If you can afford it, I got a lot of mileage out my Sennheiser MD421 when I only had a handful of mics. It sounds good on almost anything. Vocals, drums, bass… My favorite workhorse mic now is the Sennheiser e604. It's small, lightweight, and is intended for snare and toms, but to me it sounds very similar to a 421 and at a third of the price. It's fantastic on guitar amps.

Mark Rubel (MR): Probably a (Shure) SM57!

Tim Dittmar (TD): A (Shure) SM57. It's affordable, it is durable, and it works for many things. But, if you are looking to record just vocals and acoustic guitar I would purchase a large-diaphragm condenser.

John Harvey (JH): A Shure SM57 is a great choice because it is sturdy, reliable, inexpensive, and extremely versatile.

Hillary Johnson (HJ): This depends on what the application is and what you'll be using it for. Most people will tell you that you can't go wrong with a Shure SM57 and they're only around $100 and useful for dozens of things, however if you're recording vocals or something more delicate, you might prefer a condenser mic and which one would depend on your budget.

Craig Schumacher (CS): A SM57 because it's the workhorse of the biz.

Kurtis Machler (KM): Evaluate your immediate needs and your budget. A decent large-diaphragm condenser would come in handy in a variety of recording situations, but they can be a bit pricey. Condensers excel at capturing drum overheads and vocals but can be problematic in higher spl (sound pressure level) environments like close miking drums and recording guitar amplifiers. A handheld dynamic mic like the good old Shure SM57 is great for guitar amps and decent for vocals. If funds are limited, get the SM57 and start saving for a condenser. Avoid mic shopping at the mall.

Jim Wilson (JW): Well, back in the day, I would have said something like the AKG 414TLII. Reason being it is a very flexible mic with multiple polar patterns, HPF and pad. But nowadays, there are so many inexpensive condenser mics out there – so many more choices. I would suggest that you find a mic that works well for vocals, acoustic instruments, as well as capturing ambience, yet fits within your budget. A good, all-around tool like this will open your eyes to the possibilities. Just keep in mind that the "Jack-of-all-Trades" microphone is usually, like the maxim states, "Master of none."

2. What Is Your Favorite Compressor?

(MP): I don't have a favorite compressor. I like different compressors on different things. I like the DBX 160 on bass, I like the distressor on vocals. It just depends on what kind of sound I'm going for.

(**FR**): Another hard one to answer because they all sound different and they all do their own "thing." That's why studios have so many different kinds in their racks. If I had to choose just one, it would probably be a dbx 165a.

(**CS**): An Empirical Labs Distressor with the British mod as it can behave like any compressor ever built and then do some other cool tricks as well.

(**ME**): The Empirical Labs Distressor is as good an all-around compressor as I can think of. It can go from subtle to extreme, is well-made, and you will keep it forever. I bought a pair when they first came out and use them every day. There are loads of great compressors, but these do everything well.

(**JW**): I'm a bass player, so I tend to gravitate toward compressors which handle this instrument well, and have never heard anything quite as euphonic as the vintage UA 175b. They are of a very straightforward design, an early vari-mu style compressor with big, juicy transformers and tone-for-days. They are great on bass, vocals, strings as well as many other things.

(**KM**): Considering cost, function, and ease of use, I would say that the FMR Audio RNC is my favorite. I bought one when I couldn't afford anything else and still use it alongside some pretty high-end stuff.

(**GS**): All around, and probably a lot of engineers would agree, the Distressor is pretty amazing. I heard another engineer, I can't remember who, say this at a conference and it always pops into my head when I patch it in, "it's hard to make it sound bad". I think I use that as a mental crutch to justify using it so much. My new favorite, though it is only new to me, is the original Pro VLA. I picked one up cheap when they came out with the MKII a few years ago. It sat in the box for a couple of years before I ended up using it. I had used one at a studio in the past but didn't really have a chance to experiment. When I finally put mine in my rack, I was surprised by the color/character it is able to impart. You can get a really cool sound on a drum mix. I like it on bass and guitars too. At the end of the day, most compressors, when used judiciously, can be made to sound okay. Even some of the very inexpensive ones. It's nice to have options, though.

(**AM**): The Empirical Labs Distressor. I love how incredibly versatile it is – it can be very transparent or very colored depending on how you set it. My second choice would be a 1176.

(**CJ**): What day? Analog compressors all have unique characteristics. Right now, I'm in love with my Purple Audio MC77s and my Manley Vari-Mu with the T-bar mod.

(**TD**): Just like mics it depends on the situation and what sound you are picturing. I love almost any compressor. Some of my favorites include, the UREI 1176, LA-2A, dbx165a, Distressor, Tube Tech CL-1B, API 527 500 Series, Purple Audio Action 500 Series, and for an affordable option the ART PRO VLA II.

(**MR**): The UA175, a fantastic-sounding variable-mu tube compressor.

(**JH**): There are many different compressors I like for specific tasks. Empirical Labs EL-8 Distressor is the most versatile, and will work well on virtually everything.

(HJ): This depends on what I'm looking to compress or limit, so I don't have one favorite. I will use whatever is available to me. dbx makes a great line of older and newer models. The Distressor is also a handy tool. If you're looking for plug-ins, Massey's L2007 is fantastic and oh-so-simple.

3. What Is a Good Vocal Microphone?

(JW): A good vocal mic is the one which works best for the particular singer you want to capture. Some great singers don't really sound their best using the biggest and worst large-diaphragm condensers. Often, a large-diaphragm dynamic mic will work better. Much of this depends on the balance of chest voice to head voice when considering your microphone choices.

(MP): There is no one good vocal mic. I like to try several mics on the lead vocal, so I can pick one that is best suited for the quality of the voice. Often, I wind up choosing a different mic each time I have a new vocalist. Don't be afraid to try mics that are considered "guitar" mics or "bass" mics on vocals. You might be surprised to find a great vocal mic in there.

(MR): There are many, and the answer to nearly every recording and especially recording equipment question is the same: "it depends." The best vocal mic is the one that helps you achieve the sound you are striving for. You have to define what "good" means for you with a particular singer in a particular sonic and musical context. Try everything that's available to you, and figure out where you are on various spectra: from "representative" to "flavored"; from full range to bandlimited; across the spectra of different sounds and functionality. It's like everything else, you'll know when you hear it. Don't forget that under most circumstances the microphone choice is less important than the song, the performance, the singer and their comfort level, and a host of other factors. So it's best not to keep the singer waiting while you swap out mics. Put a few up, listen to them and choose, or record them all, and figure it out at a time when it doesn't impede the creative process.

(TD): I use a large-diaphragm condenser mic about 90% of the time. That being said, every singer has a unique voice and you ultimately want to have a few mics to choose from so you can A/B them and find out what suits that singer the best. You would be surprised what may end up working. It is possible your kick mic might be the mic that suits the vocal sound best or that mic you never liked before makes the voice come alive. Get to know your mics and don't be afraid to experiment.

My favorites under $300: Shure KSM27, Cascade Elroy, Cascade 731r, and Oktava 219, and for over $300: AT 4050, U47, and AKG 414.

(GS): Generally I use large-diaphragm condensers for vocals. Often one with tubes in it. Lately my go to mic that works on a wide variety of singers has been the Peluso 2247 LE. Sometimes I'll use an Audio Technica 4047/SV with great results. I'm really fond of the 4047 on a variety of sources, especially guitar amps. Sometimes, though it is perfect for the right vocalist. Dynamic mics can be great on vocals too. I tracked an album for a band where we did 21 songs in 2 days. The band was very well rehearsed and all the players were good, but we still had to be extremely efficient. For the vocals, I set up three dynamics. An RE20, an SM7, and an MD421. It was a triangle formation with all the null points in the center. The three vocalists tracked all their vocals live, together. I may have moved them around to different mics to figure which mic suited each singer best, but those were the three best dynamics in the house. Dynamics have good rejection and pick up drops off pretty rapidly once you move away from it. It worked out great in this situation. Going back to LDCs... there are new ones coming out all the time. They are getting better and cheaper. There will always be the classics, but there will always be "new classics."

(**CS**): AT 4050, for its ability to record all types of sources from the sublime to the powerful.

(**CJ**): One that suits the singer within the recorded track. That could be a blown out lapel mic or a $5000 Neumann. I tend to like my Lawson L47 tube, a Shure SM7, an AKG 414 B-ULS, or a Neumann TLM-103.

(**FR**): Any mic that sounds good on the vocalist you are trying to record.

(**HJ**): This is like asking "what is a good cuisine"? Voices are the single-most unique sound out there and every one requires trying different signal chains (starting with the microphone) to see what works best. A lot of the time it won't be the most expensive mic the studio has to offer. I've used Neumann U67's/U87's, AKG 414's, Shure SM58's, ribbon mics, everything. What's your budget?

(**KM**): It doesn't work for every voice but I love a large-diaphragm dynamic. I'm partial to the Shure SM7 right now.

(**AM**): Any microphone that sounds good on the singer you are recording. I say that slightly "tongue in cheek," but it's really true. When most people say "vocal mic," they're really thinking "large-diaphragm condenser," but that may not always be the best choice. Dynamics can sound great on vocals (Shure SM58, Shure SM7, ElectroVoice RE20, and Sennheiser 421). Then again a really good Neumann U47 (or a recreation thereof), or an AKG C12, or an ELAM 251 can truly be a thing of beauty.

(**JH**): We usually try six or eight different mics to find the best match for a particular vocalist. We often choose the AKG C-12, Neumann U-67, or Neumann U-47. These mics are all very expensive, and for someone starting out as an audio engineer I would recommend the EV RE-20 or the Shure SM7 because they sound great and they are reasonably priced.

(**ME**): Typically, people like condenser mics on vocals and you can't go wrong with general purpose mics like the AKG 414 or Blue Kiwi. You can spend a lot less and still get good sounds from the Chameleon TS-2, a great value tube mic, or some of the sE models. Consider that any microphone a person is singing into is a "vocal" mic! I love the Sennheiser 421 on vocals, and the Shure SM7. These are dynamic mics which you can use on anything. It's worth trying a few things on people because sometimes a certain voice just sounds good with a certain microphone. Use your ears, not what some ad says about what you should use. Sometimes a cheap mic from the thrift store is just the thing!

4. What Is Your Favorite Microphone? Or if You Could Only Take One Microphone With You to Record Your Favorite Artist What Would It Be?

(**MP**): If I had to pick one mic to take to the end of the earth with me to record, it would be the Gefell UM900.

(**JW**): Oooh, this is a tough one. I'm a softie for all vintage Neumann tube microphones, but if I had to pick one, it would be the unusual M269. These were a variation of the U67, but they used the brilliant AC701K tube instead of an EF86. AC701K's were also used in some of my other Neumann favorites, the KM54 and KM56 small-diaphragm condensers. The M269 excels on female voice, and is the mic I used to record my wife when we first met at a recording session years ago. It's also very quiet.

(**AM**): The Coles 4038. I am continually blown away by how good it can sound on a wide variety of instruments.

(**MR**): My Neumann U67s.

(**CS**): The AT 4050 because it is what my favorite artist – Neko Case – prefers for her vocals.

(**KM**): When I first started putting together my personal mic arsenal, CAD had just introduced their Equitek line. I got a hold of a pair of the E-100 condenser mics and out of necessity used them on everything. I would use them on the drum overheads, guitar cabs, acoustic guitars, horns, and vocals. Seriously, there are records I did during this period that the E-100 was used on every instrument. As I accumulated more mics, the CAD remained at the top of the batting order. I liked the way it sounded but I also knew how to use it on anything. For that and for simple nostalgia, I'll call the CAD E-100 my favorite.

(**FR**): My Neumann U67.

(**TD**): Whatever works for the situation at that time. Each mic has a personality so it depends what I am going for, but a U47 doesn't suck!

(**ME**): You could easily do an entire record with a Sennheiser 421. Scott Litt said that to me, and I think he's right!

(**HJ**): Probably, an AKG 414. Not only do I like the sound, but I like that it doesn't need a separate power supply. If I could only take one mic with me, I could put it in my pocket.

(**JH**): I would bring the Neumann CMV-563, a vintage tube condenser with interchangeable capsules that sounds great on everything.

(**GS**): Until I find something I like better, I guess it would be my Peluso 224LE. I've also had a Peluso 22 251 for a while, that I have only recently started to love.

I want to explore more microphones when my budget allows. But for now, I am happy with these.

(**CJ**): Beyerdynamic M160 ribbon if they are quiet. Royer R-121 if they are loud. I like ribbons. If it had to be a condenser... AKG 414 B-ULS because it is more neutral than most condensers. Neutral can be useful if you have only one mic to use. Then you shape your tone at the source – instead of with the mic.

5. Which Recording Software Do You Use? Nuendo, Pro Tools, Logic, or Another Software Program?

(**MP**): I use Pro Tools and Nuendo as my recording software. I use RMG 911 for analog recording.

(**CS**): Pro Tools unfortunately and it is a love/hate relationship for sure.

(**ME**): Pro Tools HD. All these things work well, I just got Pro Tools because it's the most common and I wanted other people to be able to tell me how it worked! This was most helpful in the early days.

(**MR**): I'm using Digital Performer now, but probably switching to Pro Tools 9 soon.

(**TD**): Doesn't matter to me, but I typically use Pro Tools or Nuendo, because that is what most studios use in my town. I have Pro Tools LE on my laptop that I use on simple, mobile projects. However, I record to analog tape 90% of the time. I use new/old Quantegy 456 2" tape.

(**CJ**): Cubase versions 4 and above use exactly the same engine as Nuendo but cost half the price. All software is buggy. I prefer analog for fidelity and reliability. Don't let all the marketing and advertising folks influence your beliefs. I also use Ableton LIVE. But I haven't heard a DAW that can sum multiple tracks better than a good analog console. I still haven't installed ProTools. In other parts of the world, Avid does not have such a strong marketing share. Do you drive a Honda, Toyota, VW, or Chevy? Oh really? You aren't professional and you are never going to get there in that vehicle. SUV = DAW.

(**AM**): ProTools.

(**FR**): Nuendo.

(**HJ**): I have used Pro Tools since the mid 90s. I've dabbled with other software but keep returning to Pro Tools due to familiarity and flexibility.

(**JH**): I have used Nuendo and Pro Tools. I prefer Nuendo because it sounds better. Pro Tools has become the most commonly used recording program, and it is helpful to be compatible with other studios and engineers.

(**KM**): Pro Tools is my preferred platform. It was the first software I learned when I made the jump from modular digital recorders to software-based systems. I put in the time to learn its flow and it works for me. Digidesign (Avid) has always been aggressive in their quest for marketplace dominance and each new version has been better than the last. I have used all of the programs mentioned and there are things about all of them that I like.

(GS): For the last few years I have been using Cubase. I came out of a studio partnership before starting up my new place and that's what we used. I also know a few other engineers who use it so it seemed like a good idea to have a network of friends all using the same stuff so we could help each other out and share projects. I deliberately stayed away from Pro Tools. I always felt like the proprietary nature of their platform did not have the user's best interest at heart. I heard horror stories of studios spending a fortune to upgrade, only to find out a month later that a next version had been released. It always seemed very expensive and limiting. Now they have opened up their platform for other hardware and I am considering switching over. There are some features I would like to see added or improved before I do it. If I do switch over, I may end up running both Cubase and Pro Tools. All the DAWs out there are amazing, complex pieces of software. They all have their flaws too. At the end of the day, it's just a tool I use to record music. It doesn't really matter which one I use.

6. Do I Have to Use a Mac for Recording?

(MP): You do not have to have a Mac to record.

(CJ): No, you don't. And I intentionally saved money by building Windows machines with Intel processors so that I could buy analog gear – gear that outlasts, outperforms any platform, and any computer. DAWs are mixing tools. Microphones & mic Pres are recording tools. Somehow the digital companies have fooled us into believing otherwise. A plugin is a postrecording tool. And it sounds equally bad on PCs and Macs.

(TD): Like I mentioned earlier, I record to tape 90% of the time. But when I do use a computer I generally use a Mac.

(FR): You may have to, but I don't.

(KM): No. You don't even have to use a computer. Personally, I'm a Mac guy but I don't believe that there is a difference in sound. A well-tweaked PC can be a beautiful thing. It's just not for me. There are way too many different manufacturers in the PC world for me to keep up with. Apple makes great products that work well with the software that I use and their customer service and tech support have always been great.

(MR): Not at all. You don't even have to use a computer for recording!

(CS): No, but it helps if you are a Mac user obviously but you can get more processing power for less money with a PC. It's all a trade-off and what works for one is not always good for another, so go with what you know and understand.

(AM): Not really.

(ME): No.

(JH): You do not need a Mac for recording.

(GS): Ha ha ha ha… absolutely not! Are they better? Maybe. I have a love/hate relationship with Macs. I ended up using PCs kind of the same way I ended up using Cubase. Early on, Macs were designed for multimedia and PCs were designed for business, so back then it actually made a difference. When I got out of college I really wanted a Mac because that's what I had been using in the computer lab and all my production labs, etc. I either couldn't afford one, or couldn't justify one, so I ended up with a PC. I was recording on ADATs at the time and mixing down to DAT. Multitrack to a computer was not really in anyone's reach at the time. Later, when I my DAT gave out and I decided to start mixing down and editing with a computer, I already had a PC, so it was natural to just get the hardware I needed for that and keep my costs low. I have always flirted the idea of switching to Mac, but also always felt a little alienated as a consumer by them.

(HJ): You can use a Mac or a Windows-based computer for recording. You can also use a multitrack tape recorder that you could find cheap on Ebay or Craigslist. Don't be afraid to experiment with your recording medium. Just don't do it on someone else's dime.

SKILLS

7. How Do I Become a Skilled Mixer?

(ME): Mainly, don't get frazzled! If you lose perspective, step away from it. Eventually you know when you're going down a blind alley, but I think when you're starting out it's easy to get frustrated quickly, then nothing sounds good! Never listen to things alone for long. Listening to one drum by itself is mostly a waste of time. To start, push up several sounds and listen to how they work together. It may be overwhelming to listen to everything at the beginning, but avoid trying to perfect one track, then another, because what matters is how the sounds work together. Think in big, bold terms! On most sounds, ±1 dB is inaudible, don't get hung up on tiny tweaks. Think about what listeners will actually notice! Always go for vivid, rather than careful. Keep doing it! You always figure out new techniques, even with the same equipment.

(MP): The only way to get good at mixing, is to mix. Experiment with different types of music and outboard gear. Trust your ears, you will get better and better.

(TD): Mix a lot! Practice and experience are essential to get good at mixing. Listen to a lot of music and analyze the sounds you hear.

(GS): Mix as much as you can. Read about mixing as much as you can. Talk to other engineers about mixing as much as you can. Experiment as much as you can. Learn to record so it mixes itself. I guess that's the real trick, but you're not always mixing your own tracks, so knowing how to deal with problems is important. Sometimes I get tracks from other engineers and my first impression is usually, "wow, that sounds good…" I always end up having work just as hard building those mixes as with my own tracks. I guess it just sounds different than my recordings, and that's refreshing to my brain and ears.

(JH): Compare your results to other recordings in the same musical genre. If your mixes don't compare favorably, try to understand what is different and make changes accordingly. It takes time and concentrated effort to train your ears, so mix as often as you can.

(AM): Mix. A lot. Every day. In as many different styles of music as you can. Also, listen to good mixes in lots of differing musical styles.

(CJ): Get good at recording the source material first. Modern mixing is often overrated and self-indulgent. Leveling and panning great tracks and great performances can be very powerful.

Many engineers believe it is their duty to touch or change every sound they record. That's not mixing. Of course, if your career is based on reinterpreting what the artist intended so that he/she can sell millions of records then your focus is beyond mixing – you become a hands-on postproduction producer at that point. And young engineers can sometimes feel entitled to screw up a band's sound because they own the DAW with the cracked plug-ins. Wrong.

So, good mixing comes from good recording. And when you screw up the track while placing the mic, or choosing the mic incorrectly, or missing the tone settings on the amp, or tuning the drums poorly, or picking the trumpet instead of the viola for that overdub, then you have to "mix around" your shortcomings. This teaches you to pay attention while tracking… and your mixes get better as a result.

(MR): Mixing is a technical art, and one gets good at it as does any artist: work, practice, study, reading, concentrated deep thinking, comparison with others' work, study of and dedication to arts of all kinds, and a commitment to learning what listening really means.

(HJ): You can get better at mixing by critically listening to records, including records you don't particularly care for, records of varying sound quality. Then, mix about five records on your own. Then master one of them. Mastering your own recordings truly helps you find your weaknesses as a mixer.

(FR): Mix a lot.

(KM): Like any other musical skill, you have to spend several thousand hours mixing music to become proficient but here are a few things that will save you some time. First, know your mixing environment and your monitors. Listen to your favorite CDs through them. Second, reference your mixes in other listening environments like your car, your home, at work, and with friends. Get a good set of headphones. This is particularly useful for checking your panning and overall stereo separation. Third, get other engineers opinions. You don't have to agree with them but it's important to hear what they have to say. Last and most important, no matter what YOU think, the mix isn't good until the client likes it. Listen to them. Learn how to get the sounds that they want out of your equipment and plug-ins.

(CS): Spend many years listening to mixes you like, learn proper unity gain structure from start to finish, understand how we perceive stereo, check your work in the real world on as many different speakers as possible, and then find a great mastering engineer and leave them some headroom to make your work sparkle and shine.

8. How Valuable Is an Education or Degree to Becoming a Successful Engineer?

(**CS**): An education can get you started properly and you can learn from others experiences and mistakes.

(**MP**): Education is always a good thing. Knowledge and experience are what help you deal with any situation you get into, and education is a big part of the knowledge side of that equation. Experience, you've just got to go out and get. However, it's easier to gain experience with something if you've got a basic understanding of how things work.

(**GS**): I believe it is paramount. Education that is… You don't necessarily have to go to school to learn it, but you have to be educated… and you have to keep educating yourself. Without the fundamentals and basic understanding of sound and its reproduction, you are like a ship adrift at sea. I am not very good at, or perhaps don't have the patience, to teach the fundamentals. I think you have to have a passion for this type of knowledge and that passion can never be taught.

(**AM**): The most valuable part of going to school to learn recording is to gain access to the equipment and to get as much hands-on time with that equipment as possible. Second is having a good foundational understanding of audio (especially signal flow) so that when you do get to work in a studio, you can contribute more to the session and not ask overly simplistic questions.

(**FR**): An education is crucial, since it helps to know what you're doing. Not sure how much a degree helps. Education comes in many forms.

(**KM**): It sure can't hurt. If nothing else, a degree shows potential employers that you are serious and can finish what you start. There WILL be dry periods where there isn't any recording work and a degree is pretty helpful in those times. There is no shame in a day job or a backup plan. Do what you have to do to get where you want to be. A formal education isn't for everyone and there are plenty of great engineers out there who have done fine without one. If you do go the academic route, chose the right program for you. Do as much research as possible before committing to a course of study. Tour the campus, read reviews, and talk to current and past students. Ask what they are doing now and if the program helped them fulfill their career goals. Be very cautious of any program that promises job placement. Know this – expensive does not always mean good.

(**JH**): A degree or certificate is useful if you want to apply for a job at a large business. A person wanting to work as a freelance engineer or small studio owner is better off spending money on equipment rather than tuition.

(**CJ**): It is important to challenge yourself. There are so many do-it-yourself engineers out there making a living and growing without formal education.

But a good learning environment can immerse you in situations that push you forward and force you into critical thinking. Some people need that. And how many engineers under 30-year-old know everything there is to know about recording? You can be an assumptive a-hole or you can be better any way possible.

(TD): As a studio owner I would hire someone with experience over someone with a degree that doesn't have experience. Someone with experience may not have a college degree on the subject but may be way more educated on recording through their experience. However, college can provide a great backup plan, educate you on the subject matter, and give you a place to meet other people like yourself to network with. Whether you go to college to educate yourself on recording or through another means, you will never know it all. That is what makes it challenging and always new to me.

(MR): It depends… These days some technical training can be helpful, mostly as a way of more quickly learning the technology and techniques. It is such a multifaceted area, and there is so much to know about, including recording and audio technology, acoustics, computers, communications, electronics, music theory and history, recording history, and endlessly onwards. The degree may or may not be important to establishing oneself, but the education never ends.

What's equally important probably is what goes along with it, mentorship and a philosophical framework. It's also very important in regards to establishing and maintaining a career, to establish a network, starting with your teachers and their contacts. Also, the peer network that is established by friendship with other students is one of the most valuable assets you will ever have.

(HJ): You don't need a degree to become a sound engineer. However, spending time studying any of your passions will help you immensely as an individual in the field of recording in many ways that are difficult to account for. My college minor was in Psychology and I can't express how helpful that has been to my recording career.

(ME): Music recording is totally word of mouth. Nobody cares where you went to school, they only care if you recorded something they like. Of course, school can make you skilled and ready to work, it may speed up things for you. Connection to the music world is mostly something you have to do by making your own way in some music scene and showing people that you have the goods. To work for a broadcast operation, it might look good to have a nice educational background. Music recording is really down-home, and totally based on grass-roots reputation, and being friends with people.

9. Do You Record/Cut With Compression or EQ?

(MP): I record with compression or EQ if it sounds good. My main goal is to capture good sounds, then I don't have to do much to them later in the mix.

(MR): I tend to record with EQ if I think it needs it, and not to track through compressors. I monitor through them if I think I'll be using them, and write down the settings if I find something nice.

(CS): Depends on the instrument being recorded and the relative abilities of the artist. I prefer to "EQ" via mic placement, choice, and preamp characteristic and "limit" as needed to prevent distortion or clipping.

(KM): Depending on the situation, I will use compression and/or EQ on the front end of a recording. If it is something that will probably need a lot of compression, like kick, snare, bass guitar, and some vocals, I like to do it in stages (a little while cutting and whatever is needed in the mix). I use front end EQ much more conservatively. On well maintained and tuned equipment, proper mic placement will usually get me in the ballpark. The artist usually knows what they want their instrument to sound like.

(ME): I don't use much compression when tracking, maybe one or two sort of special effects type things, like, say, a smashy room, but that's about it. But I don't shy away from EQ. I know I'm going to cut a lot between 300 and 600 Hz on a bass drum, so why not go ahead and do it? I know some people really do cut everything flat but I'd rather always be working toward the final tone.

(FR): Usually not with EQ, sometimes light compression. I like to keep my options open in mixing.

(CJ): Yes, Tape compression, GP9, 456, and especially ATR Magnetics formulations to 16 track 2". Maybe a little overhead compression, maybe something on the bass for control. I try to adjust the EQ at the amp or by choosing mics that shape the sound at the source. Multiple mics (in phase) can offer a way to EQ later simply by balancing levels of two different tones on the same source.

(JH): I record with EQ, compression, and effects to get the tone as close as possible to the final sound. I like to make firm decisions about sounds as I record, and then as I overdub I can tell how well the new track fits with what is already on tape. This makes mixing very easy.

(AM): Yes, sometimes, but I love it when I can cut without. Whenever possible I try to get the best microphone/preamp combination for the instrument at hand. I also try to change/modify whatever I can at the source, that is, changing guitars, pickup selection, amp settings, swapping drums (especially with snare drum), tuning drums; then moving the mic before I resort to EQ. That said, certain instruments do often need EQ when cutting, especially to get certain sounds (e.g., modern rock drum sounds – they need a lot of EQ).

In a similar vein, certain instruments are very challenging to record without compression: vocals, for example. I also (like many engineers) like to use a little compression while tracking just to add certain coloration to some sounds – for example, on bass. I might only have the compressor kicking in on a couple of dB, but running the bass through a tube compressor adds a great color.

(TD): It depends on the session. If I am cutting 10 songs and mixing them in a weekend I will cut with whatever is necessary to finish the session. Also, I tend to do more processing with novice musicians versus skilled professionals. Skilled professionals are likely to play more consistent and have better tones, meaning I don't have to apply as much compression or EQ. Exceptions would be if a unique or different sound is needed, then anything goes.

(GS): Sometimes. Depends whether it is analog or digital. Who will mix it, etc? I used to never do it because I wanted to keep my options open. Now, with DAWs, there are so many options that it can be a good thing to start making decisions early.

(HJ): Usually I track with compression or EQ, it depends on the instruments and how they are being played combined with natural room acoustics. I won't go crazy though unless we (myself and the artist/s) are 100% sure we want a specific sound. It helps to tailor your sounds but not premix them when you're recording.

10. Do I Need to Know About MIDI?

(MP): You don't need to know MIDI to record, but all knowledge is helpful in just about any endeavor.

(KM): Yes. It is part of the deal now. Don't limit yourself.

(GS): It doesn't hurt. I don't deal with it a lot. I took a class on it in college so I can muddle my way through when necessary. It depends on your workflow and what type of music you will be dealing with. Before I took a class in MIDI, or maybe it was a manual I read, I found MIDI to be completely mind boggling. The class really helped clear things up. I suppose you could automate some stuff with it, but with DAWs, automation in the box is so easy that I haven't needed it. If I'm dealing with musicians using MIDI, they are so up on it that I don't really have to get involved. I suppose I should do more with it.

(CS): Depends on the type of music you plan on making. It's the "algebra" of recording so you should at least learn what it is capable of and understand when you would need to use it.

(CJ): Yes a little. You need to know everything that your clients know. Don't waste their money. Study beforehand. Talk about the project in advance. Be respectful.

(ME): Depends on what you're doing. Dance music, hip-hop, etc. are all about it, rock music not so much. It's pretty cool, though, I wish I knew it!

(MR): Why not?

(FR): The more you know about anything to do with audio, the better your chance of getting work.

(TD): Again, everything helps, but I rarely use it with the music I record.

(JH): I usually record real instruments to 2" analog tape, and mix to ½" tape without using a computer. For my type of work MIDI is not necessary. However, it would be advantageous for someone wanting to start an audio engineering career to know MIDI. You never know when that knowledge could be a great asset.

(AM): No, but anything you can add to your "skill set" is valuable. MIDI is a tool that enables you to do certain things that can't be done otherwise. Although it's not something I use in the studio every day, when it is needed I'm very glad that I understand how to use it. Also, many people don't really understand how to use MIDI, especially when it comes to troubleshooting, so knowing how to deal with MIDI problems (connections, configurations, etc.) can definitely give you an edge over other people.

(HJ): This depends what you want to do as a career. If you are going to be doing field recording for film, no, MIDI has nothing to do with this. If you're going to be recording music, yes, MIDI may or may not be incorporated into a project you will work on. It's also important to know the basics of programs such as Reason or Ableton Live, which use some of the same principles.

11. How Do I Know When a Mix Is Done?

(CS): When it's balanced in stereo, and makes you smile on playback because it moves you emotionally and makes you want to hear it again.

(ME): When you truly enjoy listening to it! This can be difficult when you've been working for hours and are basically tired and worried! Consider that most iconic tracks were done in the days before a million revisions and recall mixes. I am still a big believer in concentrating on a mix, doing my best and leaving it alone forever. Here it is: The Mix. These days, somebody will pop-up a month after a mix is done and want to make a tiny change that nobody will ever hear. It's fairly easy to do in an in-the-box (all-computer) situation, but I think it's a bad trend. It leads to nobody ever being satisfied with anything! I think it was a truly better situation when mixing was seen more a performance, something you captured and then congratulated yourself on your genius!

No, sometimes you can redo a mix and improve it, but what I see these days is a lot of worry and the inability to get finished with anything. It's better to finish, and then make your next record!

(FR): When you and the client are both happy with it and excited about it.

(MP): I know when a mix is done when it sounds good and the client is happy.

(TD)

1. The song feels good and moves me the way I had envisioned.
2. A feeling that comes over me and I just know, or at least think I know.
3. When the budget runs out!

(MR): You feel it.

(GS): That's a good question. I'm not sure I know when a mix is done. A lot of what I'm doing when mixing is dealing with the things I don't like first. Once I get through that, then I can start to concentrate on getting everything else where I want it. I'm presented with new challenges all the time when mixing. Especially, if I am mixing another engineers tracks, or even something someone recorded at home. With DAWs I find myself going down rabbit holes of drum sound replacement, sidechaining with gates or compressors… all to try to "fix" things. When I have to do that, the mix never really sounds done.

(**KM**): The mix is never really done. There will always be someone who thinks that the bass needs to be louder (the bass player) or that the keyboards need to come down a bit (the bass player) or that the doubled tambourine part should be panned to the center (the bass player's girlfriend). The trick is to learn how to recognize diminishing returns. At some point, what seems like a good idea really isn't. This comes from practice. Also, the less people present at a mix session, the better. The best case scenario is you and one or two sober members of the band.

(**JH**): A mix is done when it either sounds good to you and your clients, or you run out of time and money.

(**HJ**): A mix is done when you can step away from it and hear everything the way it was intended to be heard… or when you run out of money ☺

(**CJ**): It makes you physically move. You might be dancing around or singing along or pounding your fist. That is important. Plus, you burn a test CD to play in your 13-year-old Honda. And you effectively measure the mix against the thousands of hours of music your brain and body have enjoyed in that familiar space. Then you stop being a stubborn jackass and let a good mastering engineer handle the final phase of the process. It's the bands' album – not yours, control freak!

INDUSTRY INSIGHTS
12. How Did You Get Started?

(**MP**): I got started when a friend built his studio. I was good at electronics and carpentry, so he asked me to help. It sounded like fun, so I said yes. Fifteen years later, I'm still here, helping out in any way I can and having a great time.

(**JH**): I had some experience doing live sound, and then went to a recording school. I got an internship at a large major-label studio in New York City that eventually turned into a job. I mopped the floor, made coffee, and ran to get lunch for more than a year before I got to assist in a recording session. I moved to Austin, Texas and opened my own studio when I couldn't find a job at an established studio. My studio has been in business for 15 years.

(**AM**): I interned at a studio in Boston while I was still in school. I don't even think I got credit for it, but I didn't care – I just wanted to be in a studio. After school I moved to San Francisco and went to every studio in town and basically kept knocking on doors and calling to work for free (interning, again). I eventually interned at four or five of the big studios out there until I got a foothold at Hyde St. Studios. I interned there for a while and proved to them that I wasn't a flake, that I didn't have an ego/attitude, that I knew my stuff and that I was willing to do whatever it would take to do the work. My first paid work there was doing the "Midnight-to-8 a.m." sessions – pretty brutal, but I did it to "pay my dues".

That progressed into more second engineering work, assisting outside engineers that would come to work at the studio. That was a great experience, working with amazing, talented folks that had been making records for 20–30 years.

(MR): I have always been rabidly interested in music and recording, and started a small studio in 1980 with a group of friends.

(ME): I played in bands, did a lot of thinking about what was actually going on in those sounds on record, and then when the Teac 4-track machines came out in the early 70s, Chris Stamey and I did nonstop recordings on a Teac 2340. That really taught me how to make things add up! We had the Teac, three dynamic mics, a little box that made the four tracks play back left, right or center, and later, an Echoplex tape echo device. Over the course of a school year we got drastically better at doing this, and we never had any EQ, compressors, or processors of any kind aside from the Echoplex. So, although this sounds like a typical old-fart story, it does prove to me that it's all about getting your hands on whatever you have and working with it. Once you have enough stuff to make a sound, you just have to go for it.

In college, I realized I really wanted to try to do recordings for real, so I sold some things my parents had given me and bought some used professional gear. This was a good move; this gear was from the days of truly "pro" gear, really high-quality, and I still use most of this stuff which I bought over 30 years ago! And I used what I learned with the Teac and just kept going with the pro gear. It sounded better and I had enough confidence from playing in bands and doing all these basement tapes that I figured the "pro" version was essentially the same task, which it is! But I think the most important thing was that I understood what the bands wanted, and tried to give it to them. I was never dogmatic about the "right" way to do things, that's boring and sort of uncool... of course, there is a bit of "right and wrong" in this work, but music recording is about fun and excitement, so bands often came to me after working in some uptight place where the engineer just yelled at them for being too loud or something. If you are working with artists, they are the #1 consideration. You can argue with them a bit if you think there's something they really need to consider, but you must always convey respect and interest in what they are saying!

(CS): By accident, being the guy who bought the gear and was curious about how to record my own band and friends music. From a garage at my house that ended up being a neighborhood nuisance to a leased space that was big enough to track bands and built up one patch cable, mic stand, guitar pick, headphone, microphone, preamp, and so on and on till you have what you think you need to do the recording you want to do. It never ends actually so I'm still getting started to this day.

(HJ): I was lucky and jumped at a chance to manage a small studio in hopes that I would train as an engineer at the same time. It was 100% about who I knew, people I became friends with in college.

(TD): Started off playing drums and guitar in my single digits. Figured out multitrack recording soon after by recording and bouncing between cassette decks. I was always the person in the band that knew how to hook up the PA. Thanks to my older musician brother. When my mom asked what I was going to do in high school I said, "surf and play drums." She suggested I have a backup plan. I found out that I could get a degree in Audio and I was sold. Eventually moved to Austin to finish my RTF degree and I was playing drums on a recording session and asked the owner of the studio if he needed an intern and he said "Yes," …25 years later still rocking and recording.

(GS): I started out as a hobbyist. I made a conscious decision not to pursue recording. I was afraid that if I recorded for a living, I would ruin something I love. It was later on down the road where I found myself in a position to pursue it professionally without that fear. I recorded a lot of bands for free or cheap. I had a studio in my house. My wife had to put up with it. Once I decided to make it my living, I invested some more money and quit doing things for free. It was pretty simple. I had an internship at a very nice studio right after college. It turned out to be a worthless experience. I don't think I was a very good intern. I didn't like the music they were doing. I kind of wasted the opportunity. But, at that age, I wasted a lot of opportunities. I had no idea where my life was going. I'm pretty happy about the way it's turned now, so far.

(KM): It was probably out of desperation but my parents actually suggested that I look into a recording school. That was good and it gave me the basic knowledge that I needed to get started. I did the usual internship, assistant thing in some studios and recorded a ton of bands on equipment that I bought or borrowed over the years. I wore out several cassette 4-tracks. That was all very important in my development as an engineer but the thing that brought me the most opportunities in studios was my ability and willingness to fix stuff. At one point, I was engineering sessions in a small record label's in-house studio at night and building patchbays and custom cables for a local pro-audio store during the day. I did that for about 2 years. The relationships that I forged doing the tech work have been some of the most important and permanent of my career.

(CJ): I actually began recording at a young age. I used portable cassette recorders to create "shows" as if I was in charge of Foley, or special FX, or general ADR (I think that is the term), by enlisting the help of all the neighborhood kids. We'd add sound to all the pages of our favorite comic books using different voices, sound effects, and musical scores. I sang in the church choir until I thought it wasn't cool. Then later in high school I guess I felt that being able to sing was cool again. I convinced some friends to let me play in their cover band. I programmed with sequencers for some of the set. My band mate and I began to record our original song ideas with a 4-track around that time and on into college.

Having to look at something two-dimensional and imagine what it should sound like, or having to reverse engineer musical structures was incredibly fun for me. I also built model airplanes as a kid too… so joining things together from small parts just seemed to be part of my nature. Plus I loved music as a kid – not because I wanted to be in a rock band – just because it seemed so amazingly interesting. So it all fits together when I look back.

Somewhere in my 20s, I decided to learn how to record and mix my own band because I wasn't very interested in what the industry was doing. I have to give credit to Fugazi and Steve Albini for helping me tie my musicianship in with my engineering skills as a way to maintain creativity. Knowing what I know and owning the gear I own simply feels empowering. And helping other musicians record is obviously fun for me if you read the first couple sentences.

So I guess I started recording folks for money by first taking all of my equipment to a remote location, recording them, then mixing at my house. Eventually I began recording out of my house much like Tim [Dittmar]. Then I began looking at local rooms/studios. Eventually I found the space I am in right now. And over the past few years I've expanded and enhanced.

So I've been recording over 20 years and I've been recording for money about 3/4ths of that time. But I think I really "got started" about 10 years ago when I began multitrack recording using: Tascam DA38s, then DAWs with 12 or more ins and outs, then finally a real analog 2" multitrack machine. That's when bands began to seek me out. So maybe that's what I consider my start. Multitrack was key… Having 16–24 live simultaneously recorded tracks all sounding good is challenging.

(**FR**): I started playing music as a kid and played around with any type of tape recording gear I could get my hands on. I learned as much as I could and read a lot of books. Then, I got lucky. I found an investor and started my own studio.

13. How Do I Get My Foot in the Door?

(**MP**): You start by recording anyone you can. You record your band, a friend's band, your neighbor's band, someone you saw play a gig, anyone. You keep recording and recording whenever you can. This will give you valuable experience, and your friends will recommend you to their friends, which will give you even more recording opportunities.

(**TD**): Skill, drive, persistence, and luck.

(**GS**): Well, you either build your own door, or just keep trying to get into any studio you can. Offer to work for free. Don't be above any job. Always be enthusiastic, and positive. Watch and learn. Ask questions. Figure out what needs to be done and do it before you are asked… even if that means taking out the trash or cleaning the toilet. If you want to build your own door, then get together as much recording gear as you can. Record a few bands for free. If those come out good, then you can start to charge a little. Put all the money into more gear. Whichever way you do it, you'll have to support yourself while you are getting things going. There are very few jobs in recording, that I know of, that pay at the entry level.

(**CS**): Be nice, humble, quiet, and eager to do menial tasks while learning the studios preferred way of tracking, overdubbing, and mixing. Pay attention and listen and show respect to all the musicians you work for and never place yourself above the artist.

(MR): Dedicate yourself to learning as much as possible, and make yourself indispensable to someone. Be the first one there and the last to leave, learn everyone's job, be a good communicator who is pleasant to be around, learn the etiquette, be humble, have a good attitude, and work hard.

(HJ): Patience is extremely essential as well as not expecting to be handed over a dream job on a plate just because you want it. This is an extremely competitive field and it takes a confident, intelligent, technically minded person who can also be willing to take a step back and not be pushy. Listening to what your prospective employer needs is essential.

(JH): Try to get an internship at a busy studio. Learn as much about recording on your own as possible. Get a small recording setup and practice. Build your resume by recording for people for free if necessary.

(KM): Your sister's boyfriend Kevin has a cousin who knows this guy at work who is friends with the drummer in a Green Day tribute band. Buy a digital 8-track recorder on Craig's list, borrow some mics and go record the band for free. It is not hard to find bands that will put up with a little inexperience for a free recording. Of course, it's all wasted time and energy if you don't learn the equipment first.

Internships are good, if you can get one. Persistence pays off. You will probably get a lot of rejections before someone actually lets you come fetch them coffee. When you do get an internship, do a good job. Stay out of the way, do what is asked of you and wait. Trust me, nobody likes a hotshot intern. Show up early, stay late, start putting mics, cables, headphones, and gear away while the band is getting rough mixes. Help the band load in and out. Learn how to solder! I'll say it again, LEARN HOW TO SOLDER. Every studio has a pile of bad cables somewhere. Fix them and you will instantly become an invaluable part of the team.

(FR): Just keep knockin'.

(ME): Get friendly with some bands and go to a session with them and tune their guitars or something. It doesn't matter what it is, but if you show that you are useful, pleasant to be around, and smart, eventually somebody will ask you to do something more substantial. If you have a recording setup, record people for free, and try to only record interesting people! Mainly, get noticed. Realize that whatever you do, it has to be truly good. You can't just coast, you have to be exceptional.

14. Are There Jobs Available in the Recording Industry?

(MP): There are very few jobs in recording that you can get in the traditional sense of going on an interview and landing a job.

(KM): It's not really a nine to five kind of gig. There are not a ton of "jobs" but there is plenty of work. Like any other freelance situation, networking and self-promotion are key factors to your success. Right now, there is a band deciding to record a CD. Your "job" at this point is to get the gig. Do good work for them and they will work with you again. And so will others. Every session that you do is an advertisement for you as an engineer.

(**FR**): Lots of them.

(**CS**): There is work but, I can't speak to jobs. Recording is about sessions and "gigs" and work comes to you based on your reputation and vibe more than your space and equipment. More and more we have to go to the work and provide our talent in nontraditional spaces so it is not so much a "job" as an occupational lifestyle.

(**ME**): Not really, in music, anyway. I suppose broadcast may still be somewhat like the old days of actual commercial activity. This is all becoming a fancy, popular hobby. You definitely can't knock on the door of a recording studio and get hired. They are mostly closing down. This is most unfortunate! Individual "star" recording people are doing OK and they may hire assistants occasionally, usually people who have already become very good at all this through their own experiences. There are jobs for hotshot editors, who can fix drum tracks, etc., on a per-song basis. Same for low-cost mastering.

(**CJ**): Yes, if you are passionate and you serve musicians fearlessly – you will stay employed.

(**TD**): Definitely, but you have to be extremely driven and motivated. You will also need to be creative and have other skills (booking, web design, guitar instructor, sandwich artist, etc.) that you can utilize if you don't have a lot of steady recording session work.

(**GS**): The best job in recording is the job you create for yourself. Sure, there are internships and the hope of landing some sort of recording gig, and I suppose some people take that route, but for me, it's all about controlling my own destiny. Even if it's not true, I believe that I am in charge of my own destiny and I can say "no" if I don't want to work on a project.

(**JH**): It is difficult to get a job at a recording studio. The number of recording schools has increased in the last 20 years, so there are many graduates looking for jobs. Simultaneously, the number of large commercial studios has decreased due to sluggish CD sales and increased competition from smaller studios that are staffed only by the studio owner. It is possible to create your own job by working as a freelance engineer or starting a small studio business yourself. This is more difficult than collecting a steady paycheck from larger company, and it requires a great deal of motivation and perseverance to succeed.

(**AM**): Yes, but not in the traditional sense of working for someone or a business. Studios don't employ engineers – engineers are freelance agents. So it's a position wherein you need to find your own work.

(**MR**): Yes, everywhere and in all kind of unexpected ways.

(**HJ**): There are always jobs in recording. The amount of pay is the catch. No matter how much experience you have, you will always be tempted to work for free. Unless they are signed, decent bands usually have no money. It's still fairly easy to get an unpaid internship in a bigger recording studio and work your way up but it can take time. There are no quick paths to making your dream happen. But enjoy the work you do and do your best at all times. This is what opens up future possibilities of better pay.

15. How Much Money Should I Expect to Make?

(MR): You can make hundreds of dollars a year doing this crap! But seriously, it's highly variable, depending on the situation.

(CS): Hopefully enough to support your gear habit and the bill collectors satisfied. Beyond that, just working making music and recording is a reward in of itself.

(CJ): More than me I hope. Figure out what you are good at – what you do differently than other engineers/producers – then strive to be the best in that area. There's room at the top. And the old farts are going to go deaf or retire eventually. Make good artists shine and bad artists happy and you will make money.

(MP): I'd love to be able to say you get rich recording, but you don't. If your goal is to get rich, you need to move on to another line of work and record as a hobby.

(KM): I think a reasonable expectation for an entry level position in the recording industry would be zero dollars. If you work hard and have a bit of luck, you could eventually make millions. The good news is this, if you choose recording as a career because it is something you love doing and you go into it with sensible goals and work hard, the money will find you. The longer you continue to do good work and improve your technical and interpersonal skills the less competition you will have.

(ME): None, unless you get famous with something people really like. If you record a band who get noticed, other people will want you to record them, and you might be able to shake a little money out of the situation. Really, the money is a tiny fraction of what it was 20 years ago. Very few bands are making money, so they can't pay much, indie labels are usually tiny, etc. Money is only available when a lot of people know who you are and think you will really make great records. Good people do get noticed!

(TD): You may work for free a lot. You will find that most really good bands, unless they already have backing, don't have any money. If you want to work with such a band you will often weigh how much you really want to work on the record versus how much money you will take at a minimum. This happens all the time! Don't go into audio engineering because you want to be rich! Like any profession, the better you are the more likely you are to get paid for your talents.

(AM): Not a lot, especially at first.

(FR): Expect to make millions, but don't be disappointed with hundreds.

(JH): Making records is generally a low-paying job, so you have to want do this type of work for reasons other than making lots of money. There are some engineers who are paid more than the industry average because they have already made many commercially successful records. These people are in the minority of working recording engineers, and it took them years of hard work to get there.

(HJ): This depends on where you live and what you want to do. You could land a paid internship at a post house and be able to pay your rent, or you could get an unpaid internship at a local recording studio and do that for a year or two.

(GS): Depends what you want to record. I think there are people out there making a lot of money and they believe they are making important recordings. For me, it's about working on projects that are satisfying to me personally. I feel like I've spent more on recording than I've made… that's not true, but if you own your own studio it can be very expensive. Freelancing can work out if you can build up a client base and/or your reputation. Look on the Internet at studio rates. This might give you a sense of what you can make. These rates will change over time and they vary from studio to studio.

16. What Do You Think the Future of Recording Will Be?

(MP): I think the future of recording will be the same as it is right now… Trying to capture the essence of someone's art with an ever-changing array of tools and media.

(AM): I have no idea…

(ME): Hard to say! I miss the world of high quality, expensive rock records by obscure but good artists! There will probably always be a small "pro" scene for mega stars and everybody else will be recording on home setups and getting variable results. It's sort of like professional photo portraits giving way to snapshots. Of course, smart people will do good things no matter what the situation is. But I lament the decline of a world of music recording experts who make a decent living and have tremendous experience and expertise.

(CS): Hopefully we will see higher bit depth digital recording, perhaps a better algorithm than PCM encode/decode and recognizing that being able to reproduce frequencies above 20k actually matters.

(MR): It will be as varied, creative and marvelous as the musicians and music that we serve.

(TD): It will probably be an extension of what it is today, constantly recombining old and new technologies. People will continue to be creative using Apple iPads, 4-track cassettes, laptops, and whatever inspires, is affordable, and is available to that particular person.

(FR): One day, you will be able to think up a song in your head and instantly it will be live on the internet, ready for anyone to download it straight from your brain.

(JH): Computer recording systems will become more powerful and hopefully improve in sound quality. I think the number of very large commercial studios will continue to decrease as computers improve. There will always be people who want to get together in the same room and record the music that they create.

(**HJ**): The future of recording is unknown. With the push toward lesser-quality, compressed files, it's our responsibility as audio people to insist on better standards, that's as far into the future as I can see.

(**KM**): The future is now. Who knows what the next recording trend will be? I don't. I'm pretty sure that as long as people are listening to and making music there will always be a need for talented engineers. Learn signal flow, mic placement, EQ, compression, and practice mixing. These skills will always be at the heart of the recording process. Keep an open mind and be ready for whatever comes next.

(**GS**): Recording technology continually becomes more attainable for the average musician. Lots of bands are recording themselves. Many artists use the computer as part of their creative process…recording what they need and composing in the box. I see more of this in the future, but the recording studio will never go away. There will always be people who want someone else to engineer and produce for them. There will always be people who are more inclined to record and produce than write and perform music. It's a natural partnership.

(**CJ**): Awesome. I believe digital's role is to catch up to analog. And since analog is still amazing, the future will be more of the same for those of us who can hear the difference. Sadly, when digital finally comes around…only the analog recordings of days past will be able to make the conversion. You can't upconvert an MP3. You can't upconvert a CD, or even a DVD. I don't get the trend right now. When you go to a movie and you hear the music and sound FX are so rich and full with so much information for your brain to enjoy… how can you go home and put on an MP3? And if you are a musician… how could you put up with the quality of your music being destroyed… why have ears?

Yes, current digital technology makes editing easier. But PCM (pulse code modulation) technology doesn't sound good. Do some research. Trust your ears. But computers are really a joke. Two dimensional representation of the real word experience is just silly. Let's design a computer and DAW that can maintain the fidelity of analog and place the engineer or musician into a 3D space – instead of ones that force us to propagate antiquated technology born out of a savvy R&D division of a consumer electronics company.

I am optimistic that eventually computers will be so powerful that digital file compression is unnecessary. Then artists will be deciding how their music will be delivered. They will also choose the extent of low end and top end in their recordings. It won't be decided by Sony Phillips or Apple. Theatres have gone to 70 mm, IMAX, IMAX 3D and now the home theatre systems are following. But we still can't find files better than mp3s quality on portable devices, or even in our cars. It's just digital information. Stop controlling it. And stop letting them control it.

I want to hear something that can't be reproduced on vinyl, CD, DVD, or Blu ray. Something that literally breaks the sonic mold. I can't wait till [sic] the day when the only way you can hear a certain album is when the artist allows you to do it. And the experience is so unique and mind blowing that people travel great distances to hear it. I hear music like this in my dreams. And I am sure others do as well. Who cares if you can have 40,000 songs on an iPod if you can't make it to that once in a lifetime musical event?

That is exciting. Maybe the new technology will surpass analog pathways as well: Laser microphones, magnetic resonance imaging, who knows. But even in that future, you still need a good room and a good engineer AT THE BEGINNING to get those sounds into the new medium. A good room makes the source sound cooler. A good engineer – like a good photographer – can artistically frame the shot, set the tone, and make the subject "larger than life."

The future of recording is forever loud and clear.

IT TAKES TIME BUT IT'S WORTH IT!

It will take many years of hard work and dedication to succeed in the extremely competitive field of audio engineering. The rewards will be many. I love it when I finally get a copy of a CD or record that I put my heart and soul into. To me, I get way more satisfaction having completed and been part of a music recording than from getting paid. The money helps pay the bills but the tangible product of what I accomplished is much greater.

Although you can make a decent living as an audio engineer you shouldn't enter the field expecting to make large sums of cash right away. Experience means a lot with music production. So go get some. Learn as much as you can anyway you can. Record with whatever you have at your disposal. Take into account, you will need to be very determined and patient at the same time. Audio engineering can be a fun hobby or one of the coolest jobs you could ever imagine!

BALANCED VERSUS UNBALANCED

If you become an audio engineer, sooner or later you will recognize differences between balanced and unbalanced audio equipment. A balanced line offers excellent immunity from hum, noise pickup, interference, and ground loops. Most professional recording studios use all balanced equipments. An XLR or a ¼" T/R/S (tip/ring/sleeve) connector usually indicates a balanced input or output (see Figures in Appendix B). A balanced line is used with microphones and high-quality line signals.

The Technical Part Is as Follows

Balanced lines typically have THREE wires used for a single signal. Pin 1 is usually ground, Pin 2 is HOT (+), and Pin 3 is COLD (−). A balanced line indicates that the signal-carrying wires are at equal potential, but opposite polarity. Balanced lines and cables are typically used in professional settings and can be run very long lengths without accumulating noise.

A RCA connector or ¼" T/S (tip/sleeve) usually indicates an unbalanced input or output. ¼" unbalanced lines are common with instruments such as electric guitar and bass. RCA, or phono, is typically used with record players, speakers, and consumer stereo systems. This type of connector is also used to connect audio and composite video to TVs. They are often color-coded red for the right speaker and white for the left speaker. Unbalanced lines need to be kept short to avoid picking up interference or other unwanted noise. They typically are not effective at lengths over 15'.

The Technical Part Is as Follows

An unbalanced line has TWO wires, a HOT and a GROUND. The tip is HOT (+), whereas the sleeve is ground.

▲ TIP

Some rack-mounted signal processors will allow you to switch between −10 and +4. This option is typically found on the back of the piece of equipment. If your studio is set up to be balanced, make sure that +4 is selected on all equipments. Select −10 if all your recording gear is unbalanced.

MORE ON DECIBELS

Up to this point, we know that amplitude is measured in dBs. However, as you gain more experience in the coming years, you will come to know that there are many types of decibel ratings. dBs can describe power or voltage and sound pressure differences. Consumer and professional equipment not only differ in cost and perceived quality but also differ in the actual audio signal levels. Professional audio equipment generally has a normal operating level of +4 dBu, whereas consumer audio has a normal operating level of −10 dBV. Notice the change from dBu to dBV? Professional audio is about 12 dB hotter than consumer equipment. This difference in level can degrade audio quality in two ways:

1. If the output of the pro gear is plugged into the input of the consumer gear, the higher levels may overdrive the input, causing distortion.
2. If the output of the consumer gear is plugged into the input of the pro gear, the signal will be lower, which results in an overall lower signal to noise ratio and the input signal may be barely audible.

Decibels

dBFS – decibels at full scale. A dB rating associated with digital audio where zero is the maximum level. **−20 dBFS = 0 VU = +4 dBu** This is not a fixed standard.

dBu – professional rating +4 dBu = 0 VU = 1.23 Volts RMS and is used to express voltage levels.

dBV – consumer rating −10 dBV = 0 VU = 0.316 Volts RMS and is also used to express voltage levels.

dBSPL – decibels sound pressure level. Measures sound pressure levels relevant in our daily lives. This is the dB reference used mainly throughout the guide. 85 dBSPL = 0 VU = +4 dBu.

Appendix B

AUDIO CONNECTORS

A variety of connectors are used in music production. It will be helpful if you can recognize the differences and uses of each connector. A jack is used to describe the female side of an audio connector, whereas a plug is used to describe the male side of a connector. A plug is inserted into a jack. As mentioned in Appendix A, some connectors are associated with balanced lines/signals, whereas others are associated with unbalanced lines/signals. In addition, some connectors can be used for both analog and digital purposes. Let's start with analog connectors.

Analog Connectors

Four common connectors associated with analog equipment are as follows:

1. PHONE (aka 1/4", T/S, or T/R/S): Originally adopted by Ma Bell for use with phone operator switchboards. Commonly used with instruments (guitar, keyboards, and bass) and patchbays. Comes in both T/S (tip/sleeve) and T/R/S (tip/ring/sleeve). A T/S will have a single band around the tip of the connector, whereas a T/R/S will have two bands around the tip.

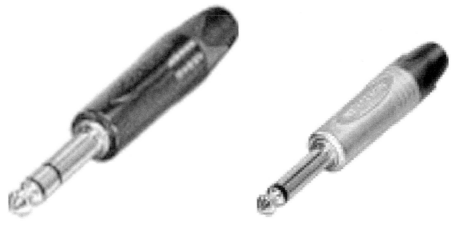

Figure B.1

2. XLR (aka Cannon, Mic, or Neutrik): Typically a three-pin connection but is available in other pin configurations. Provides a very secure connection. These connectors are usually associated with mic cables and balanced audio. The male end of the connector is most often used for input, whereas the female end of the connector is used for output.

Figure B.2

3. MINI (aka 1/8", 3.5mm): Smaller version of the 1/4". Comes in both T/S and T/R/S. Associated with personal stereo headphones, iPods, and computer sound cards.

Figure B.3

4. RCA (aka phono): Usually comes in mono pairs and is used in conjunction with consumer audio/stereo setups. Record players generally have RCA outputs. The tip caries the signal and the ring is the ground. This is an unbalanced connector.

Figure B.4

 TIPS

It is extremely crucial to maintain the integrity of digital audio. Digital signals use many of the same type connectors as analog; however, they are usually higher quality and gold tipped. A S/PDIF can use unbalanced RCA phono type connectors, whereas AES/EBU can use three-pin XLR balanced connectors.

A few other connectors associated with analog equipment are as follows:

Tuchel: Come in various pin configurations. They have an outer locking ring and are well known throughout Europe. Smaller tuchel connectors are common with some vintage mics, whereas larger tuchel connectors may be found on some older model recording consoles and high-fidelity audio equipment.

Figure B.5

ELCO/Edac: Multipin connector used in conjunction with tape machines, consoles, and sound reinforcement.

Figure B.6

Banana plug: Used to connect higher end audio amplifiers and speakers.

Figure B.7

Speakon: Designed as a tight fitting speaker connector. Excellent choice with live sound reinforcement.

Figure B.8

Digital Audio Connectors

AES/EBU: Created by the Audio Engineering Society and European Broadcast Union to develop a standard when digitizing analog audio. The connector looks like a typical XLR and is often used with a digital work clock and/or I/Os. Other connectors may be used for AES/EBU protocol such as a DB25.

Figure B.9

S/PDIF (Sony/Phillips Digital Interconnect Format): An audio protocol associated with digital consumer audio gear. Used to interconnect components over shorter distances. The connector looks like a common RCA connector but can also be found as a TOSLINK connector.

Figure B.10

TDIF (TASCAM Digital Interconnect Format): It is a proprietary connector used by TASCAM. TDIF uses a bidirectional connection. Unlike the ADAT light pipe connection, it can both send and receive up to eight channels of information with a single cable.

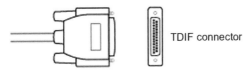

TDIF connector

Figure B.11

ADAT Lightpipe: Developed by Alesis to carry up to eight channels of uncompressed audio at 24 bit and 48 kHz using fiber optic cables.

Figure B.12

Optical/TOSLINK: Usually associated with an optical connection. Newer equipment often uses a TOSLINK, replacing the coaxial S/PDIF connector.

Figure B.13

MADI (Multi Audio Digital Interface): Another audio protocol besides S/PDIF and AES/EBU used to transmit digital audio. MADI supports a greater number of channels. AMS Neve, Soundcraft, Lynx, Avid Technology, and many other audio companies use MADI protocol. Typically uses an optical connector.

USB (Universal Serial Bus): Designed for communication and as a power supply for computers and other electronic devices. Can be identified by its trident logo. New versions have been made available that significantly increase the speed at which data is transferred. The connector comes in several versions. Pictured here are the common Type A and Type B. USB Flash drives are often used to backup and store important files.

Figure B.14

Fire wire (Apple's high-speed serial bus design): Often used to transfer data to external hard drives, webcams, and Apple's own iPod. Firewire is capable of transferring audio and video at about the same speed as USB 2.0.

Figure B.15

DIN MIDI connector: A five-pin round connector used for MIDI protocol. Preceded USB and Firewire. Used to interface synthesizers, lights, and other various controllers.

Figure B.16

DB25: A 25-multipin connector. Used with analog or digital audio signals. Often used with analog for an eight-channel snake or other multichannel audio configuration. With digital audio, DB25 is commonly used for AES/EBU multichannel signal. DB25 connectors are great for a tidy and secure connection. However, the pins are easily bent and they are very difficult to repair and build.

Figure B.17

A few links to purchase audio connectors, cables, and parts are the following:

www.partsexpress.com

www.markertek.com

www.redco.com

Analog A sound recorded and reproduced as voltage levels that change continuously over time, such as with a cassette tape, vinyl record, or analog recorder.

A/D Analog to digital. An audio interface or other A/D converter can provide an analog to digital conversion when needed. The A/D converter changes an analog signal to a digital signal.

Click Often used to establish a consistent tempo when recording basic or rhythm tracks; referred to as a metronome.

Cochlea The snail-like structure of the inner ear that detects sound pressure changes.

Comb filter An effect or acoustics issue which results in a series of very deep notches, or dips, in a sound or rooms frequency response.

Compressed The reduction of an audio signal's dynamic range or file size. In the digital world, an audio file can be compressed to save storage space, as with an MP3.

Control surface Allows the user to turn knobs and faders instead of clicking a mouse. Looks like a small mixing board; used in conjunction with a DAW.

DAW Digital audio workstation; used to describe a digital audio set-up as opposed to an analog set-up.

dBFS Decibels relative to Full Scale; measurement of audio levels in a digital system. 0 dBFS is the maximum level in the digital world. Go over zero and the waveform will be clipped resulting in an unpleasant sounding and typically unusable signal.

De-essing(-er) A plugin or rack mounted signal processor that helps remove unwanted sibilance ("s" and "sh" sounds); usually applied to a vocal track.

Digital A means of encoding data through the use of the binary number system. Digital audio isn't continuous in that samples of the sound are taken and reconstructed to appear like a sine wave. Digital audio examples include CDs and MP3s.

Distortion Occurs when a signal is too strong or too hot for a particular audio device. It is part of the sound and is not considered noise. Many guitar players regularly use distortion as a part of their sound.

Duplication The process of burning or duplicating audio or data to a blank disc; what many people identify as CD-Rs; great for demos, rough mixes, and short-run CD projects.

Feedback A loop that occurs between the input and output of a signal; can occur if you point a mic at a speaker or a guitar at an amplifier.

Fletcher-Munson curve A group of subjective plots and curves that measure the ear's average sensitivity to various frequencies at different amplitudes. In the early 1930s, Fletcher and Munson came up with this hearing test. According to the Fletcher Munson curve, the average human ear hears best between 1.5 kHz and 4 kHz, especially at lower volumes. At about 85 dB, the average human ear hears all frequencies as equally as we are ever going to hear them without affecting the perceived pitch.

Flutter echo A term used for an acoustic problem caused by parallel walls and surfaces. If you clap your hands and hear a quickly repeating metallic-like sound, you know that the room exhibits flutter echo. Diffusion and other acoustic methods can be used to eliminate or control this issue.

Haas effect A psychoacoustic phenomenon that occurs when we can't distinguish the direction of a single sound panned one way when combined with a second delayed signal panned the opposite direction. The Haas effect occurs when the delayed signal is under about 20 ms.

High-cut or low-pass filter (LPF) A button or switch often located on a console, preamp, or mic that, when selected, cuts high frequencies and passes low frequencies at a predetermined frequency setting. A high cut can be effective on bass, kick drum, and other instruments that do not need extreme highs.

Ips Inches per second; professional analog recorders generally record and playback audio at 15 ips and 30 ips. At 15 ips, a standard reel of 2" tape yields about 33 minutes of recording time. At 30 ips, a standard reel of 2" tape yields about 16 ½ minutes of recording time. Recording at 15 ips will provide a better bass response but with increased tape hiss. 15 ips is better for louder music. Recording at 30 ips provides a tighter low-end and a decrease in noise. 30 ips is great for quiet and acoustic styles of music.

In-line console(s) A mixing console that allows a single channel to control both input and monitor signal paths. Many newer analog and digital mixers are in-line consoles.

In the box Phrase that indicates all music production is completed in a DAW, or computer.

I/O Input(s) and output(s). Located on audio interfaces, recorders, and mixing boards.

ISRC International Standard Recording Code (ISRC) is the internationally recognized identification tool for sound and music video recordings. An ISRC exclusively and permanently identifies the recording to which it is assigned regardless of the format. The ISRC code contains the song title, country it was recorded in, record label information, and track number.

Jitter A time-based error that can occur during the analog to digital conversion or when more than one digital device is hooked up to another.

Low-cut or high-pass filter (HPF) A button or switch often located on a console, preamp, or mic that, when selected, cuts low frequencies and passes high frequencies at a predetermined frequency setting. A low cut can be effective on some vocals, cymbals, electric guitar, or other instruments that do not need extremely low frequencies.

Low-frequency oscillator (LFO) A control that sweeps through a signal's low-frequency range to create vibrato, tremolo, or other pulsing/rhythmic FX. A common control on synthesizers and effects pedals.

Masking A subjective phenomenon where one sound appears to hide another sound. This is common with two like instruments that occupy the same frequency range as with two electric guitars.

Midfield monitors Monitors typically used in larger control rooms that provide a louder, fuller sound; larger than common nearfield monitors.

Mix The result of blending instruments, sounds, and voices typically to 2-tracks/stereo.

Mbox A digital audio interface common with Pro Tools.

Mono or monaural A sound system with only one channel, regardless of the number of speakers used.

Oscilloscope A test instrument that allows you to see the size and shape of a sound wave.

Pan(ning) A knob, pot, or other control that allows a sound to be placed anywhere in a mix.

Pot Short for potentiometer; pots are knobs on a console or other audio component used for a variety of controls such as panning or gain adjustments.

Preamp An electronic amplifier typically used to increase the volume of an audio signal.

Producer Person in charge of the creative side of a session, but can also be a technical person. Unlike an engineer, a producer may get involved in the song writing, song selection, and other internal aspects of the music. Often by default, an engineer may take on some roles of a producer.

Protocol A special set of rules defined for a specific use.

Punch (-in and -out) A punch-in or punch-out is the process of adding or deleting a part to a predetermined section of a recording. This process involves going into and out of record mode and is typically used to correct a part or add/overdub additional sounds.

Quantization Represents the amplitude component of digital audio; the higher the bit resolution, the smoother the digital representation of an analog signal.

Red Book The standard protocol for audio CDs. The Red Book contains the technical specifications for all CD and CD-ROM formats. It is named after a series of books bound in different colors. The Red Book standard for a CD is 44.1 kHz and 16 bit.

Replication The process of manufacturing professional CDs or DVDs. Think of this as the professional version of a duplicated CD. Replicated CDs are purchased through a disc manufacturer and usually require minimum order of 300 discs.

Rpm Rotations per minute; determines the speed a vinyl record is meant to be played.

Sibilance An unwanted "s" or "sh" sound in the 4 kHz–10 kHz range that often accompanies a vocal; can be eliminated with a de-esser or other EQ methods.

Sine wave The simplest waveform in audio; consists of a single frequency and has a musical pitch, but a neutral timbre because no harmonics exist.

Split console Unlike an in-line console, a split console has only one fader or monitor option per channel. A console is often "split" during recording, with half the channels acting as signal outputs to the recorder (sends) and half the channels acting as monitors (returns).

Standing wave Acoustic phenomenon that may be created by parallel walls and surfaces. A standing wave is produced by the interference and reflection of sound waves and is determined by dividing the speed of sound by 2(L). These waves are commonly referred to as room modes.

Stem Grouping or sub-mixing instruments or sounds. Stems may include grouping drums and bass together, grouping vocals together, or grouping any like sounds into a mono or stereo submix.

Stereo or stereophonic Refers to a sound or system that involves two separate channels. These channels are labeled left and right (L/R). A standard audio CD is stereo, containing 2-tracks, left and right.

Synthesis Associated with electronic music; occurs when a sound is created without an acoustic source (ex. synthesizer).

Take A recorded performance of a song; often described as a good or a bad take.

Tweeter The smallest speaker in a typical speaker or studio monitor that reproduces higher frequencies.

Woofer The larger of the two speakers on a typical studio monitor. The woofer reproduces lower frequencies below a certain frequency range.

Index

Page numbers followed by *f* indicates a figure and *t* indicates a table.

Credits

Video 1 Music by Murdocks, Die Together, 2008
Video 2 Music by Starry Eyed, Head High in Hawaii, 2005
Video 3 Music by Boonesboro, My Dog Loves You, 2011
Video 4 Music by Dead Waiter, 2008
Video 5 Music by Ross Rossman, 2011
Video 6 Music by annabella, Sun is King, 2008
Video 7 Music by The Hearts and The Minds, Bending Trees, 2010
Video 8 Music by the Buzzkillers, For the Kids, 2005

Videos – Produced by Andrew Miller
Talent – Landry Gideon
Sound and Mixing – Tim Dittmar
Photography – Andrew Miller, Tim Dittmar, Anne C. Kelley

Audio Clips
1.0 Music by annabella, Peachtree, 2008
3.1 Music by Ross Rossman, "Great White Killa," 2011
3.2 Bass, Joshua Zarbo
3.3 Guitar, Tim Dittmar
3.4 Drums, Tim Dittmar
3.5 Music by bo bud greene, "Heads and Friends," 1995
3.6 Music by annabella, "La Ciudad," 2005
3.7 Music by Kristi Rae, "Any Life At All," 2006
5.0–5.3 Flute, Terri Dittmar
7.0/7.1/7.2 Drums, Tim Dittmar
7.2/7.3 Vocals, Terri Dittmar
7.3 Electric guitar, Andy Bracht
7.4 Guitar Tim Dittmar
7.5 Strings by The Apple Trio, 2011
7.6 Guitar, James Currey (the Buzzkillers 2009)
7.7 Guitar, James Currey (the Buzzkillers 2009)
7.8 Strings by The Apple Trio, 2011
7.9 Bass, TJ Smyrson (the Buzzkillers 2009)

Illustrations by Anderson Bracht

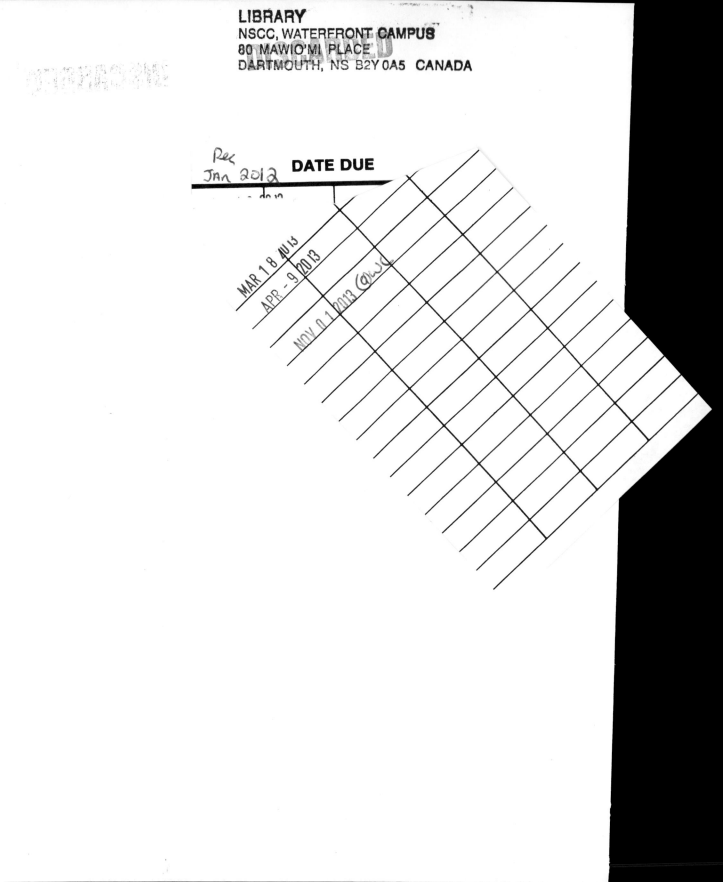

DATE DUE

Rec
Jan 2012

MAR 1 8 2013

APR - 9 2013

NOV 0 1 2013